"A down-to-earth description of how to run efficient and successful M&A and how to avoid failures."
Johan Molin, CEO, AssaAbloy

"The basic idea of *The M&A Formula* is very interesting. Learning from successful investors in corporate M&A, the patterns and processes the authors use in applying them to their own acquisition plans is a fascinating concept. If a successful investor would not invest in an acquisition, it is probably a good idea for a corporation not to follow such a plan and vice versa."
Günther N. Fuhry, Executive Vice President, Swarovski

"*The M&A formula,* from my experience, hits the nail on the head: simple principles with easy-to-use toolkits reflected against the theory. A must-read for all M&A professionals."
Stefan Schneider, M&A Integration Expert, Editor in Chief M&A Review

"*The M&A formula* provides a useful, hands-on framework for successful M&A with numerous real life examples. A must-read for all those fascinated by M&A."
Daniel Couvreur, Conference Board Director, M&A Council

"The authors present a comprehensive framework for achieving M&A success, and illustrate its value through a detailed and very insightful analysis of how the best and the brightest achieve sustainable success with M&A."
Niels Bjorn Andersen, Professor Emeritus, Copenhagen Business School

"The headline 'the opportunity, the pain and the promise' describes the content at its best. The 50% failure rate of M&A deals must encourage any person in charge to wake up. This book woke me up, and made me change my views on M&A deals."
Roland Wilcke, Officer with procuration, Günzburger Steigtechnik GmbH

"M&A is a combination of art and science. Advice on getting it right pays huge dividends in achieving the intended outcome of the deal, and *The M&A Formula* gets it right - it's a good book."
Jacqueline D. Reses, Capital Lead and People Lead (CHRO), Square

"*The M&A Formula* provides a uniquely insightful and compelling analysis of a longstanding conundrum in mergers and acquisitions. In the M&A domain, the persistent popularity of the transactions counterbalanced against the often devastating destruction of shareholder value has puzzled investors for decades. Based on their extensive experience, the authors provide, for the first time, a formula drawing on large-scale bundled data analysis."

"Proposing the innovative concept of TRA (total return analysis), *The M&A Formula* assesses the impact of a bundled set of M&A interrelated activities on corporate performance. Secher and Horley break through the logjam of analyzing individual acquisitions within short-term time horizons, to instead take a crucial longer term and programmatic perspective. Their findings, observations and recommendations are key, as M&A transactions are not isolated occurrences but rather systemic phenomena. With a fresh perspective, *The M&A Formula* provides a way forward for scholars and practitioners alike—highly recommended."

Professor Kathleen Marshall Park, MIT Sloan School of Management

"A contemporary approach to the classic challenge of creating value from mergers and acquisitions."

Professor, Dean Annette L. Ranft, North Carolina State University

"*The M&A Formula* has a singular purpose: to help managers improve their odds of better acquisition outcomes. The authors detail a sensible and experience-based three-step formula for M&A success. Systematic tools are explained for evaluating targets, managing the process, and dealing with external advisors and building internal M&A competencies. These tools are then carefully explained with case applications. I found their claims that every firm has an M&A reputation that can be used as a tool particularly insightful. And their frank and honest treatment of the role of financial advisors in M&A was refreshing. This book is a must-read for executives interested in improving their firms' M&A outcomes."

Professor Bruce Lamont, Florida State University

"For me as a business student, *The M&A Formula* sheds light on a continuously growing sector and provides a very practical point of view on tactics that are successful in the real world."

Wolfgang Höck, Student, Strategic Management & Law, MCI Management Center Innsbruck

"A must-read for anyone who is looking for a new model for successful M&A based on an outstanding combination of research and practical applications."

Lena Scholz, Student, Strategic Management & Law, MCI Management Center Innsbruck

The M&A Formula

Proven Tactics and Tools to Accelerate Your Business Growth

Peter Zink Secher
Ian Horley

WILEY

This edition first published 2018

© 2018 Fixcorp

Registered office

John Wiley & Sons Ltd, The Atrium, Southern Gate, Chichester, West Sussex, PO19 8SQ, United Kingdom

For details of our global editorial offices, for customer services and for information about how to apply for permission to reuse the copyright material in this book please see our website at www.wiley.com.

All rights reserved. No part of this publication may be reproduced, stored in a retrieval system, or transmitted, in any form or by any means, electronic, mechanical, photocopying, recording or otherwise, except as permitted by the UK Copyright, Designs and Patents Act 1988, without the prior permission of the publisher.

Wiley publishes in a variety of print and electronic formats and by print-on-demand. Some material included with standard print versions of this book may not be included in e-books or in print-on-demand. If this book refers to media such as a CD or DVD that is not included in the version you purchased, you may download this material at http://booksupport.wiley.com. For more information about Wiley products, visit www.wiley.com.

Designations used by companies to distinguish their products are often claimed as trademarks. All brand names and product names used in this book are trade names, service marks, trademarks or registered trademarks of their respective owners. The publisher is not associated with any product or vendor mentioned in this book.

Limit of Liability/Disclaimer of Warranty: While the publisher and author have used their best efforts in preparing this book, they make no representations or warranties with the respect to the accuracy or completeness of the contents of this book and specifically disclaim any implied warranties of merchantability or fitness for a particular purpose. It is sold on the understanding that the publisher is not engaged in rendering professional services and neither the publisher nor the author shall be liable for damages arising herefrom. If professional advice or other expert assistance is required, the services of a competent professional should be sought.

Library of Congress Cataloging-in-Publication Data

Names: Secher, Peter Zink, 1967– author. | Horley, Ian, 1965– author.
Title: The m&a formula : proven tactics and tools to accelerate your business
 growth / Peter Zink Secher, Ian Horley.
Other titles: The m & a formula
Description: Chichester, UK ; Hoboken, NJ : John Wiley & Sons, 2017. |
 Includes index. |
Identifiers: LCCN 2017037698 (print) | LCCN 2017045093 (ebook) | ISBN 9781119397922 (pdf) |
 ISBN 9781119397946 (epub) | ISBN 9781119397960 (cloth)
Subjects: LCSH: Consolidation and merger of corporations.
Classification: LCC HD2746.5 (ebook) | LCC HD2746.5 .S425 2017 (print) |
 DDC 658.1/62—dc23
LC record available at https://lccn.loc.gov/2017037698

10 9 8 7 6 5 4 3 2 1

Cover Design: Wiley
Cover Image: © billyfoto / iStockphoto

Set in 12/16pt AGaramondPro by SPi Global, Chennai, India

Printed in Great Britain by TJ International Ltd, Padstow, Cornwall, UK

For everyone who believes that there is a better way to do Corporate M&A

CONTENTS

Acknowledgments — *xi*
About the Authors — *xiii*
Introduction: The Opportunity, the Pain, and the Promise — *1*
Why Do M&A?
What Will You Learn From This Book?
Who Is This Book For?
Why Was This Book Written?
How Was This Book Written?
Who Wrote This Book?
The Professors, the Case Insights, and the Research Insights

Part I THE M&A FORMULA — 29

1 *What Is the M&A Formula?* — *31*
Before the M&A Formula …
Should You Build or Buy?
Understanding Business Model-Driven M&A
Success and Failure in M&A

vii

2 Be Business Model Driven (Step 1): What Is Business Model-Driven M&A? — 51

Business Model Complementarity and Signals for Success
Three Real Business Models: M&A Drivers and Complementarity Score
What Are the Goldman Gates?
How to Create a Target List Based on the Goldman Gates?

3 Communicative Leadership (Step 2) — 69

Lesson #1 in M&A Leadership: Drive Hard with Soft Management Tools
Lesson #2 in M&A Leadership: Create a Strong Foundation for Your People
Lesson #3 in M&A Leadership: Silent People Are NOT Team Players
The M&A Deal Committee
Intent-Based Leadership

4 Take Ownership (Step 3) — 91

The M&A Playbook
Governance in Corporate Processes
CFA: Company-Specific M&A Engagement Letters
M&A Launchpad
Funding and Corporate Finance

5 Accelerating Your M&A Formula: Digitizing M&A — 117

Strengthening the Formula
The M&A Dashboard
Digitizing: The Human Side of M&A
Now, Science Meets Business

Contents ix

Part II THE M&A FORMULA APPLIED   131

6 Case Insights (CI) and Research Insights (RI)   133

CI1: DSV: From Ten Trucks to €10bn
CI2: RB: 'King of OTC' Improved Value Proposition and Client Relation as Business Model Driver in M&A
CI3: Heritage Comes First at LVMH
CI4: The Global Brewer: Driving M&A Growth One Beer at a Time
CI5: Danaher: The Importance of Teamwork
CI6: FrieslandCampina: A Merger of Equals or an Impossible Utopia?
CI7: ASSA ABLOY: The Highest Total Return to Shareholders in Our Research Period (1st Jan 2007–1st Jan 2017)

7 Next Steps   201

M&A Reputation as a Tool for Success?
Can You Build a Good M&A Reputation and Attract Sellers?
Things to Avoid Internally—Bias
Where to Get External Advice?
FON-A-PAL—Building Your Resource Library
What Do We Mean When We Talk About External Advisors?
Financial Advisors/Investment Banking
M&A Myth-Busting
What Your Success Looks Like?

Index for Case Insights & Research Insights   227

Index   232

ACKNOWLEDGMENTS

In creating this book, we had enthusiastic meetings, interviews, feedback, and content from so many talented people. We are extremely grateful to all of them.

For the Research Insights that support the underlying formula for this book, our thanks to 'the professors,' whose years of research were summarized and applied to the Case Insights that give this book credibility and actionable advice. They are: Dr. Florian Bauer, Professor at Leopold-Franzens University Innsbruck; Dr. Svante Schriber, Associate Professor at University of Stockholm; Dr. David R. King, Associate Professor at Iowa University; and Dr. Kurt Matzler, Professor at University Bozen, Italy.

We appreciate enthusiastic leadership and innovation modeling from: Gareth Garvey; Alan Plaughman; former captain and bestselling author David Marquet. Data Collection and Research, Erik Elgersma (author: *The Strategic Analysis Cycle*) and Mads Jensen, Jensen Capital Management.

For the valuable Case Insights we thank our M&A panel, who were transparent and enthusiastic in opening up their experiences for us to share with you. Of the dozens of contributors, we especially thank: Cees 't Hart, the former CEO of FrieslandCampina; Jan Barsballe of ASSA ABLOY; Marc Koster from Heineken; Kurt Larsen of DSV; John Schultz of Verizon; Klaas Springer, treasurer at FrieslandCampina; and Johan Molin, CEO of ASSA ABLOY.

Finally, for the heavy lifting and support in editing, curation, and overall craft, we give special thanks to our writing partner and overall word guru, Kathryn Gaw, as well as the team at our publisher, Wiley.

Thank you everyone!

ABOUT THE AUTHORS

Ian Horley is an author, entrepreneur, and marketing strategy consultant for many SMEs. Ian's focus is on designing 'product market fit' programs to help merging companies grow better, faster, and cheaper.

Ian is a public speaker and a mentor to entrepreneurs, students, non-profit organizations, and educational institutions in Europe and North America. Ian's aim is to develop the M&A Formula into a movement targeted at transforming the careers and fortunes of the individuals involved in M&A, and is currently preparing the companion course at www.themandaformula.com to expand the mission's reach.

Ian Horley

Peter Zink Secher is a former banker and corporate finance specialist with 25 years of experience on M&A sell-side. Peter moved to M&A buy-side as he started working for a former M&A client some 5 years ago. He has been involved in M&A projects in most parts of the world and is now running his own company; Peter's business motto is "half of your M&A success are all the deals you don't do."

Peter Zink Secher

(*continued*)

(*continued*)

One of Peter's most important decisions in life was to take about 18 months out to devote time to his father and brother, who both tragically died of cancer in 2014. Spending time in a hospice made him understand that terminally ill people do not talk about work. They speak about the things that matter: family, hobbies, special occasions, and achievements. Make sure you achieve something in life—make a difference. Peter has published an article, and given speeches, about the terminal cancer stage.

INTRODUCTION: THE OPPORTUNITY, THE PAIN, AND THE PROMISE

The normal failure rate of at least 50% in corporate M&A is unacceptable. Trillions of dollars are wasted every year in failed M&A deals that should never have been considered in the first place. This high failure rate destroys small and medium businesses, devalues big corporates, and ruins careers.

So why do so many highly educated, bright, and well-paid people spend so much time on corporate M&A when the outcome would be the same if they just flipped a coin? Why do so many students spend years at university, only to end up in a career where there is at most a 50% chance of success?[1]

This question has been bothering us for many years, as we struggled to understand how any CEO or any owner of a small or medium-sized enterprise (SME) could allow these high failure rates to stand. During Peter's corporate tenure at FrieslandCampina, there was not one single failed M&A transaction; no acquisition was ever regretted and every deal was reasonable in line with synergies set at the binding offer stage.

By this point, Peter had more than 25 years' experience working in corporate M&A (both buy-side and sell-side), and he realized that a pattern was starting to emerge. He began to interview M&A peers in companies which had a history of making close to 500 successful acquisitions, asking each one the same three questions:

- What are your business model drivers in M&A?
- What are your main processes?
- How have you managed to avoid the normal failure rate of at least 50%?

Their responses were illuminating, and confirmed his suspicion that there must be a formula for M&A success. A formula that will make the at least 50% failure rate a myth.

That M&A success was not just a matter of luck. It was down to the implementation of a formula that focused on the individual targets of each potential deal, involved the whole organization, and put in place a number of processes which could be carried out in-house to speed up each transaction.

This is the M&A Formula. It has been based on an exhaustive study, backed up with the academic research and real-world examples that you'll see in this book. This is the M&A Formula for success, which has been created, tested, and shared with you.

The M&A Formula

1. Follow business model-driven M&A.

2. Strong leadership and communication.

3. Take ownership.

The implementation of this formula is laid out in Figure I.1.

Knowingly or unknowingly, this formula has been used by some of the most successful names in corporate M&A, creating their M&A success, and increasing value for their shareholders.

Introduction

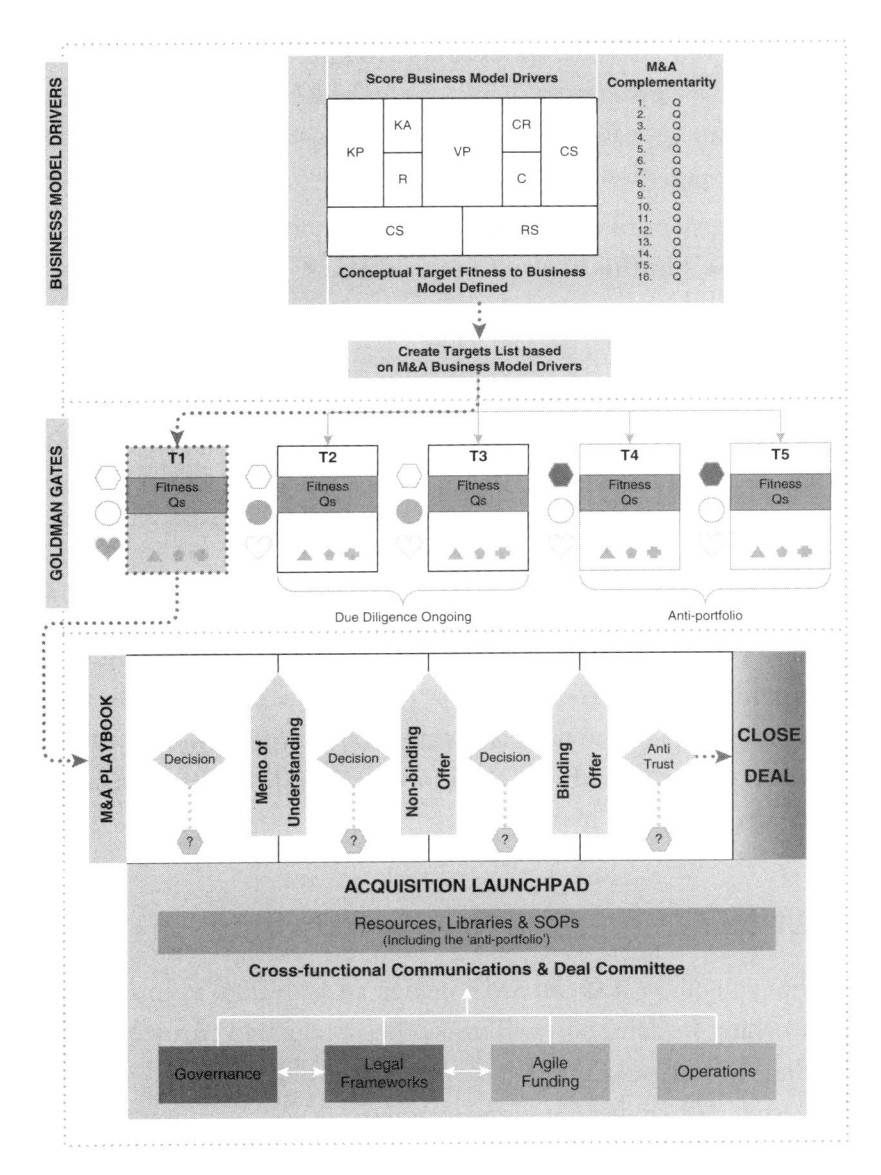

Figure I.1 The M&A Formula (see more on www.fixcorp.co)

Through our research, we were able to find countless examples where deals have progressed thanks to the M&A Formula, and many more examples of deals that failed because they didn't take these three rules into the equation.

The three steps of the M&A Formula may not make sense to you right now, but this book will explain exactly what is involved at every stage, and what to do once you have run the formula and started making deals.

We will do this by using real-world examples and academic research, as well as anecdotes from Peter's own career. We hope that by sharing this formula, we can help to reduce the global M&A failure rates, and empower more businesses to use M&A to their advantage.

Here's why the formula presented in this book is so exciting. Whilst we reference large blue-chip corporates, the formula has been proven to work for businesses of all sizes, especially SMEs. So now you can skip the expensive trial-and-error process that has dashed so many M&A aspirations in the past, and embark on the growing of your business or career, armed with the know-how and tools that were only available to those with global wallets.

Why Do M&A?

Whether you are a CEO, an SME owner, an ambitious employee, or a hard-working student, you will know that if you aren't growing, you are falling behind, and the rate at which your universe is growing is accelerating. When deciding how to accelerate your corporate growth, you are faced with a choice to build or buy, and like never before, the choice is more often to buy because we will show you how to be successful in corporate M&A.

And it is a lot simpler than you think.

When Peter became Head of Corporate M&A at FrieslandCampina in 2011, he already had 25 years of corporate and investment banking experience. Yet he had never heard anyone explain in one short sentence why FrieslandCampina was actually doing M&A. He challenged the organization to come up with a simple statement in answer to the question: "Why do we do M&A?"

Peter himself grew up on a farm, in a farming family, and clearly understood what a high milk price meant to a farmer.

In this case, the milk price was just another way to define the return to shareholders—it could be return on equity (earnings per share according to Peter), but for the farmers who held the majority of shares in FrieslandCampina, the milk price was paramount.

When people make statements like "we could build a stronghold in South-East Africa" as a way of justifying a deal, this is not a business model driver. In other words, it's not going to increase the 'milk price' (aka your shareholder value).

Before any deal, you have to be able to stand up and state what's in it for the owner. Why do we do M&A? If you don't have a clear reason—a reason that will increase your 'milk price'—then you have probably just figured out why your M&A failure rate is so high.

High failure rates are likely to be an issue in global M&A for the foreseeable future, and it is these rates which will separate the great from the average. There is no international ranking for corporate M&A, but we do have industry data which clearly shows which companies are creating value for their shareholders in corporate M&A, and which ones may as well be flipping a coin.

In corporate M&A there is a lot of room for mediocre performance; after all, who will know? Meanwhile the external advisors will pocket some more money and by the time a deal starts to fail, it's too late to do anything about it, and the C-suite pays no matter how involved they were in the deal. This is why every business should take a hands-on approach to M&A developments, and you should treat every single deal as if your career (or your wallet) depends on it—because it just might!

This book will prove the importance of understanding and planning for M&A, regardless of whether you are the CEO or an intern. When everyone understands why they are doing M&A, all the pieces of the puzzle will start to fall into place and success will follow.

What Will You Learn From This Book?

In this book, we wish to share and explain the M&A Formula. The formula offers a route to M&A success, for any organization of any size.

First, an organization must clearly identify why M&A should be part of its business—this will involve identifying clear business model drivers. We call this 'Business Model-Driven M&A.' In this book, we will share Case Insights, which are relevant to each of the most common business model drivers in corporate M&A, and we will present global companies who were able to use these drivers to create M&A success. We will also show you Research Insights from some of the world's leading academics who evaluated these companies. These insights cover both global corporates and SMEs, and we were able to compare the firms and conclude that the formula for M&A success is the same.

Next, we will reveal how corporates of all sizes can further increase their chances of M&A success by creating a 'sense of belonging' in the organization—the importance of leadership and communication based on a clear mission: business model-driven M&A.

Finally, we will show you that the way to M&A success has to come from within the organization—by taking ownership of your own transaction instead of relying solely on external M&A advisors.

The first thing you have to do is to stop doing the wrong M&A deals. If there's one thing that is clear from our research and experience, it's that half of M&A success lies in the deals you don't do.

But you will never hear this from your closest M&A advisors. Just imagine what would happen to the total M&A fees earned by investment banks, legal advisors, and transaction service providers (e.g. accountants) if all companies stopped doing the wrong M&A deals.

External advisors cannot possibly solve the problem of high M&A failure rates, because they simply don't have a vested interest in saying no.

That doesn't mean that there's no place for external advisors in M&A—quite the contrary. But when it comes to lowering failure rates, the change needs to come from within.

You have to have that moment of transformation, where you decide to use M&A for success.

After all, it is your company at stake here. What CEO would want to make any M&A-related decisions if the expected outcome was worse than a 50/50 bet? M&A-savvy corporates know why they do M&A, and how each deal fits into their business model from end to end. They know how to optimize their intake of external resources, but they also know when to step away from a deal that is destined to fail.

We will take a closer look at the relationship between corporates and external advisors later in this book, but for now let us go back to the high failure rates and why they prevail.

Let's compare direct investments such as corporate M&A with traditional fund management, which is just another type of (indirect) investment.

As in corporate M&A, all fund managers will tell you they are the best, but the truth is that very few of them are actually any good, and the rest are simply relying on luck. A truly great investment manager will be easy to identify: they will probably be ranked on the famous 'Citywire 1000' list (which identifies the top 1000 fund managers in the world); they will have fund fact sheets showing their average return on a 1-year, 3-year, or 5-year basis; or maybe they will simply be recommended by a trusted associate.

Unfortunately, there is no like-for-like chart when it comes to direct investments in M&A, but we wondered what it took to rank as one of the world's top fund managers, and whether the M&A world could learn anything from the equally competitive world of investment management. One of the best-known global rankings for fund managers is the Citywire 1000 'World's Top Fund Managers' list.[2] We took a deep dive into the investment processes of the world's top 10 investors and even had an interview with one of them. In fact, this wasn't much of a challenge, as Peter is married to Rikke—one of the top 10 world-ranked investors. We asked her: "What does it take to become top in your field?"

She said: "I'm not sure ... I just do the same stuff every day. I invest the money. Look for value."

Can it really be that simple?

'Value' is at the core of every single sector in the business world, but it can be difficult to master. It is not just about finding the best deal; it is about understanding the risk so that you can avoid failure.

This is just as important in corporate M&A as it is in fund management, but it is often overlooked in favor of closing the deal.

That doesn't mean that you can eliminate risk altogether—after all, risk is often unpredictable. But you can eliminate those risks which are obvious. For instance, Peter's wife (Rikke) revealed that one of her tried and tested investment strategies is to selectively buy corporate bonds right after an announced M&A transaction. This is because the capital markets are acutely aware of the risky nature of M&A transactions and the high failure rates, so it is possible to pick up a bargain while the market is cautious. This was a strong indication that some companies apparently have a formula for M&A success. After all, she would not have become No. 7 globally (and No. 1 in Europe) by buying assets from M&A-savvy corporates if they often failed in M&A, would she?

This will not be news to seasoned M&A professionals. McKinsey has published an article about this phenomenon entitled 'Managing the market's reaction to M&A deals,' in which the authors (Werner Rehm and Andy West) claim that announcement effects are a good instant measure of market sentiment but a poor indicator of longer-term value creation. According to Rikke, she will be 'RISK ON' when others get fearsome and the majority turns to 'RISK OFF,' but only with companies who have a string of successful M&A activities behind them. "I want the same CEO and team to do it again," she added.

A strong history of M&A success will always catch the eye of top investors, gradually strengthening your company over time and creating new opportunities for investment.

We studied the other top 10 fund managers in the Citywire 1000, and found that most of them shared the same recipe for success: dedicated research, strict and formal investment process, long-term focus, sticking to your strategy, knowing when to get out, avoiding shockers, taking a structured approach, studying competitor behavior, global diversification, bottom-up approach, and analytically driven processes.

Successful investing in corporate M&A is not entirely dissimilar. It is about following certain procedures, your 'dos and don'ts,' and sticking with your plan. We call this the 'M&A Formula,' and used correctly it will form the foundation of your company's M&A approach, allowing you to repeat your successes again and again.

After learning more about Rikke's investment strategies, we decided to observe the winning behavior of another woman in her field of expertise: Serena Williams. As one of the most successful female athletes in the world, she certainly knows a thing or two about minimizing failure and repeating her successes over and over again.

What interested us about Williams was not necessarily how she played the game (although, of course, her games are phenomenal to watch), but how she behaved before she started playing. Before the deal negotiations even start.

What we saw was a top sports person entering the battleground with her headphones on. Although she greeted the audience, she was focused on the loud music in her ears which was setting her up for battle. She warms up in the same way every time, always doing the same thing over and over again. This is a professional support system which she has built up with her coach, so that she can run through her pre-match preparation with minimal distractions, and minimal risk.

This is what it takes to become a winner, and to keep on winning time and time again.

The best people in M&A have their own 'pre-match' routines. And just like the top 10 investment managers, and the world-famous sports stars, these routines are all about reducing uncertainty on an M&A deal. The top M&A professionals can take out your bad shots and institutionalize M&A processes so that your failure rates stay low.

Unfortunately, there is no global scoring board or Grand Slam equivalent for successful M&A professionals, which makes it harder to find the right people for the right jobs. That is one of the reasons why high failure rates will prevail in corporate M&A globally.

The M&A panel behind this book are all seasoned practitioners and even though they do not have a global ranking, their results speak for themselves. We hope to inspire you with our own tales of success and failure, and advice on how to get ahead in this complex world.

Who Is This Book For?

Are you an executive with a winning mindset, a passionate SME owner, or an ambitious employee who wants to take control of your future? Do you want immediate improvements and the ability to execute M&A success again and again?

Then this book is for you.

For too long, M&A failure rates have stymied good businesses and prevented exceptional people from reaching their full potential.

M&A is an efficient way of growing or expanding any business, no matter the size. Done correctly, it is a great way of achieving (and even exceeding) your professional goals. In this book, we will show you real-world examples of transformative M&A—from the delivery firm that went from ten trucks to almost €10bn in revenues, to the locksmith who more than tripled its value in 10 years by conducting smart M&A.

Despite the fact that the global statistics are akin to flipping a coin, luck plays no part in M&A. It is all about meticulous planning and an understanding of the key ingredients of M&A success. Make your own organization punch above its weight and not rely on externals.

We have interviewed the world's most successful M&A corporates, as well as academics, investors, and analysts, and we have discovered that every M&A success story shares a number of striking similarities. Each one of these companies pursues business model-driven M&A. And each one of these companies exceeds shareholder expectations time and time again.

The M&A Formula has a proven success rate, and it can be applied to any M&A transaction of any size. Whether you are a family-run business, a rising star of the tech world, a global corporate, or a would-be M&A leader, you will never achieve greatness without understanding these rules.

Why Was This Book Written?

The world of M&A is often misunderstood, yet mergers and acquisitions are embedded functions of all corporate activities. Even if you decide that your company is not going to take part in M&A, you could find yourself becoming an M&A target. M&A is everywhere.

Corporate leaders are constantly looking for ways to improve performance and with the rise of 'financial sponsors' (private equity, activist investors, hedge funds), the pressure is higher than ever. New clients, more effective distribution, better relations, higher quality, and faster innovation are constantly being demanded, and M&A is the best way to get hold of new skills, technologies, products, and IT quickly and effectively. Still, the majority of these deals go wrong because of a bad business model fit.

The Formula for Bad M&A

1. It is not clear what you want to do with the business model.
2. Your own people do not understand what you want.
3. Many companies listen too much to external advisors, who will pocket their success fees regardless of whether or not the deal is a success.

This book is for anyone who wants to have M&A success. It will give you the ability to immediately lower the failure rate and reduce uncertainty in M&A transactions by drawing on the experience and advice of industry leaders who have successfully merged organizations and/or made great corporate acquisitions that have created value for everyone involved.

How Was This Book Written?

The team behind this book does not believe in learning from failure when it comes to corporate M&A. With a very positive mindset, we therefore decided to learn from success and not stand in line with all the others pointing their fingers at corporate M&A. Some academics, businesspeople, and investors truly believe that M&A cannot be justified at all. We tend to agree, so long as we only look at the average global failure rate of at least 50%. But this book is not about average companies.

We wanted to write about M&A success—how any company in the world can learn from the companies with the highest achievable M&A success, whilst at the same time being acknowledged by global leading academics for their M&A Formula.

Would it be possible to find real-life case studies with proven M&A success? Would these companies actually be able to explain their recipe for success? And would they be likely to repeat this success several times, while also having a high impact on the acquiring firms or the merged entities?

We tried to filter out any kind of noise and focus on M&A drivers which demonstrated a long-term better performance that was not impacted in the short term by externalities such as raw material prices, unusual sector performance, changes in legislation, country risk, stock market volatility, etc. In short, companies that were not affected by what is referred to as idiosyncratic risk or simply VUCA (volatility, uncertainty, complexity, and ambiguity of general conditions and situations).

Here is one practical example of noise from real life that was never going to be used in this book. Some years ago, one of the companies in our M&A Elite, ASSA ABLOY, bought the lock company Yale. The day after the announcement, shares of ASSA ABLOY went up by 16%. This is highly unusual, but it looked like ASSA ABLOY had just picked up a real bargain.

However, what happened with the general stock index in the UK and Sweden in the aftermath of the announcement? Did the

British Pound appreciate against the Swedish Kroner, and were there expectations of more currency movements? Could you possibly isolate the effect? For instance, what if investors got a little overexcited that day, and one or two weeks later the hype wore off and ASSA ABLOY started trading at a more reasonable level? Or was the immediate increase in share price caused by a hedge fund in a short squeeze on Yale shares?

So-called 'M&A Event Studies' investigate the share price of the buyer and bidder companies around the announcement day, or a short period after announcement. But most of this global research is rubbish, as there is simply too much noise embedded in short-sighted research.

We started to analyze firms left, right, and center, but we didn't really get anywhere, until we took our research away from the Internet and moved the discussion out into real life, with real people in real places. We spoke to professional money managers, the shareholders of global firms who managed real-money accounts with a long-term horizon of at least 5 to 10 years (avoiding hedge funds and private equity firms). We wanted face-time with the 'alpha' investors who have a proven track record of success. As such, we avoided 'beta' investors and passive funds, which are just tracking an index of some kind.

We asked these investors if they knew of any stock-listed global corporate which was highly active in M&A and had delivered superior returns for them in the past 10 years.

We didn't have a preference for stock-listed firms over any other kind of ownership. We also didn't like global corporates more than SMEs, or any other sized firms. Nor did we like active fund managers more than passive ones—our choice was merely an academic necessity.

Firstly, to reduce as much 'noise' as possible, we defined a research period from January 1st, 2007 to December 31st, 2016. The so-called 'fear index' VIX is the popular measure of implied volatility of S&P500 Index options, as calculated by the Chicago Board Options

Exchange (CBOE). The fear index of the stock market has been on a long-term average of about 15–18%. When the index is trading lower, it's an indication that the market is calm. By the end of our research period, it was trading close to an all-time low of 10–12%.

At the start of our research period, the VIX index was trading at its highest fear factor in 30 years, hovering above 50% during the 2008 financial crisis. In some periods during our 10 years of research, the fear index has been above 40% (2009 and 2010) and twice it has been above 25% (2012 and 2015).

Another member of our M&A Elite, Louis Vuitton Moët Hennessy (LVMH), went from a share price of €42 to €123 in less than a year after the 2008 financial crisis. Amazingly, luxury goods were selling better after the crisis.

From a theoretical point of view, you have to separate organic growth from acquired/merged growth, but in practice that would be like trying to separate hot and cold water from a bath tub.

When we speak to global corporates who are highly active in M&A, we are reminded of this point over and over again. Already, 12–18 months after an acquisition, things are becoming one entity without the ability to separate running the business from buying businesses. That's why we only wanted to focus on serial acquirers who have M&A as an embedded corporate growth engine over a defined period of 10 years.

The way we have defined M&A success is by one metric: total return analysis (TRA).

Any such analysis could be misleading if a short-term horizon was at hand, or if a company had only carried out a few M&A transactions with very low impact. A company like Apple has been very successful in terms of delivering a high total return to its shareholders alongside other firms. The FAANG stocks (Facebook, Apple, Amazon, Netflix, and Google) delivered an index growth from 100 (5 years ago) to almost 500 (at the end of our research period), but the total return created for shareholders cannot be attributed to M&A growth, merely a (highly successful) organic growth. You can be successful without doing M&A, but in relation to this book it proves nothing—we prove that M&A can be a value-adding growth driver alongside organic growth, contrary to the global M&A failure rate.

Total return analysis (TRA) is the amount earned from a shareholding over a specific period (when dividends to the shareholder are reinvested in the shares) + capital appreciation (increase in the stock price itself).

TRA is also referred to as the 'true growth' over time, meaning the 'big picture' of a company's historical performance. The TRA approach with a 10-year long horizon was perhaps the single most important factor in our analysis. The most prominent school of thought in M&A derives from financial economics deeply rooted in the performance of stock market-based measures. Still, it can only be done on stock-listed firms and in our view only with a long time horizon. Yet, we can still learn from the Global M&A Elite.

However, other schools of thought must be taken into consideration when analyzing non-listed firms such as SMEs. Firstly, there are no available capital market prices available for the majority of SME firms. Secondly, finding a number of firms which are as representative of SME M&A success as the Global M&A Elite was practically impossible. That was the main reason why we teamed up with leading academics to test the M&A Formula on SMEs. This book presents a highly representative match between large corporate and SME M&A research and their way to M&A success. 'The Antecedents of M&A Success' is based on a sample of 106 SME transactions from a publication in the *Strategic Management Journal*.

This was as close as we could get to a like-for-like comparison with listed global corporate results.

However, we didn't want to limit our research to online tools and Internet references. We conducted several calls, joined research and planned meetings with the leading academics behind this great research, and all parties only identified a growing piece of evidence, that the M&A-savvy practitioners and the world's leading academics had finally found something that could be the formula of M&A success. Not a guarantee of course, but a framework which, if followed and executed the right way, should produce a higher chance of M&A success, making the at least 50% failure rate a myth—at least for those who want success.

We asked the academic team for an exclusive latest update on 'The Antecedents of M&A Success.' We knew that the companies we had analyzed in the Global M&A Elite would definitely do more M&A deals, but what about the SME firms? Our academic team put together a questionnaire and asked the SME firms: "M&A: would you do it again?"

Professor Dr. Florian Bauer told us that, based on a study conducted in 2017 with 113 respondents: "45.1% of SME firms either agree or strongly agree that they would pursue more M&A deals in the future. However, 56.6% of all SME respondents agreed that the acquisitions of the past five years have contributed significantly to corporate success."

We had not only concluded that the Global M&A Elite was able to create success by following the M&A Formula, but that SME firms were able to do it too (see Table I.1).

The two frameworks, 'Business Model-Driven M&A' and 'The Antecedents of M&A Success,' were based on companies of totally different sizes. The research methodology was—and had to be—totally different as, for instance, capital market values do not prevail for the majority of SMEs.

Table I.1 The M&A performance miracle

Research Insight title	The M&A performance miracle
Case Insight to discuss	This book does not investigate the performance of an individual acquisition, but rather the contribution of a bundle of coupled M&A activities on corporate performance. This seems to be a highly relevant issue.
Context from Peter Secher	We wanted to kill the myth that the at least 50% failure rate applies in any given case.
Research Insight content	**Point:** Stock market data reflects on the expectations of shareholders and is often associated with the real and true value. Similarly, accounting-based measures seem to be accurate and objective. **Counterpoint:** Stock market data is highly sensitive to the environment and internal developments. Furthermore, it is a unidimensional measure for a multidimensional concept. Accounting-based performance measures lack comparability across various countries (due to different valuation standards), and they reflect on the past and not the future. There is empirical evidence that short- and long-term stock market, as well as accounting-based, measures share only little variance. As the performance of acquisitions is highly motive-specific, research often suggests that managerial assessments (e.g. with surveys) are superior. **Contingency:** M&A is a strategic pathway for corporate development and thus, it is not a single acquisition that matters, rather a bundle of coupled M&A activities that contributes to corporate performance.

Table I.1

References	
	Point:
	Zollo, M., & Meider, D. (2008) What is M&A performance? *Academy of Management Perspectives*, August: 55–77.
	Counterpoint:
	Cording, M., Christmann, P., & Weigelt, C. (2010) Measuring theoretically complex constructs: The case of acquisition performance. *Strategic Organization*, 8(1): 11–41.
	Contingency:
	Almor, R., Tarba, S.Y., & Margalit, A. (2014) Maturing, technology-based, born-global companies: Surviving through mergers and acquisitions. *Management International Review*, 54: 421–444.
	Meglio, O., & Risberg, A. (2011) The (mis)measurement of M&A performance—a systematic narrative literature review. *Scandinavian Journal of Management*, 27: 418–433.

Let us take a closer look at other types of methodology in the world of M&A.

There are basically four different schools of thought in M&A academia. We have based all Case Insights on the 'Financial Economic' school, which is based on the wealth effect for shareholders as measured by TRA. Most academic literature and equity research analysts base their conclusions around announcement day, whereas we base it on a 10-year timeframe to reduce as much noise as possible.

The 'Strategic Management' school centers around the 'strategic fit,' which has some conceptual similarities to business model-driven M&A but is less *specific* in describing the M&A drivers for success. It is more holistic and focuses on similarities and complementarity in various aspects. As it often involves non-listed firms, the way to measure M&A success can be supplied by accounting-based values as opposed to market values like TRA.

The 'Organizational Behavior' school is the third type of measurement in M&A success/failure. It focuses mainly on cultural integration and sometimes on task integration. Do the organizations work well together? What is the degree of integration?

The final school is the 'Process' school, which has a distinct focus on the speed of integration and other topics like communication (see Table I.2).

Inasmuch as we like all frameworks, we kept our focus on the 'Financial Economic' school as it uses real market values. Very few conclusions can be challenged when using a long-term horizon, and we can all learn from the world's largest companies which have seen repeat success in M&A.

Who Are the Global M&A Elite?

Historical performance is no guarantee of future performance in any of the M&A Elite companies we have analyzed. However, we do emphasize that their M&A actions throughout the research period are indisputably value-adding, and only this period should be used as an example.

For instance, we know that RB acquired Mead Johnson (the baby food manufacturer) after our 10-year research period had ended. This acquisition looks to be largely in line with most of RB's takeovers, but it is a slightly different deal, in terms of the business model building blocks. The channels, or routes to market, for the baby food sector are quite different to the over-the-counter (OTC) channels which RB normally masters. RB is sometimes called 'The King of OTC,' but baby food is sold via different channels to the other products in the firm.

Table I.2 Four schools of thought in M&A research

Research Insight title	The four schools of thought in M&A research
Case Insight to discuss	This book draws an interdisciplinary and integrative perspective on M&A.
Context from Peter Secher	The Case Insights in this book are based on the 'Financial Economic' school using a TRA with a 10-year time horizon of 2007 to 2017. However, other generic examples of M&A success are shown, drawing on the three other schools of thought in M&A.
Research Insight content	**Point:** The four schools of thought reduce the complexity of the research field and thus allow for an in-depth analysis and understanding. **Counterpoint:** The fragmented and specialized background of M&A research strongly limits the development of a more holistic understanding of performance antecedents and consequences. **Contingency:** M&A success depends on pre-merger as well as post-merger issues and consequently, it is about the interrelationships between the (commonly individually analyzed) success factors.
References	**Point:** Haspeslagh, P.C., & Jemison, D.B. (1991) *Managing Acquisitions: Creating value through corporate renewal*, Free Press, New York. **Counterpoint:** Cartwright, S., & Schoenberg, R. (2006) Thirty years of mergers and acquisitions research: Recent advances and future opportunities. *British Journal of Management*, 17: 1–5.

(continued)

Table I.2 *(continued)*

Contingency: Bauer, F., & Matzler, K. (2014) Antecedents of M&A success: The role of strategic complementarity, cultural fit, and degree and speed of integration. *Strategic Management Journal*, 35(2): 269–291.

We have therefore not included any M&A deals from January 1st, 2017 for any of the firms (see Table I.3).

The Global M&A Elite has consistently returned value for their shareholders. If you had invested €100 in any one of these firms, you would have made a return of about 300–400% (although ASSA ABLOY returned 427%). In the past 5 years, we also happen to know that close to 60% of SME firms have experienced corporate M&A success. We just didn't know how they did it. At least not until this book was written.

Who Wrote This Book?

> *Don't you think it's strange that some of the most well-educated and highest paid bankers, lawyers and accountants on this planet are delivering M&A sell-side advisory services to corporates when the expected outcome is like flipping a coin?*
>
> IAN HORLEY, SME OWNER AND CO-AUTHOR

Two years after Ian asked Peter this question, this book was created. At that point, Peter had just taken a long break from the world of M&A, after 30 years working as both a sell-side advisor/banker and a corporate buy-side professional. The rest of this section is in Peter's own words.

Table I.3 The Global M&A Elite

Bloomberg Ticker	Revenues (€bn)	Market Cap (€bn)	Est. P/E Dec. 2017	Website	Industry	Employees
DSV DC Equity	9.1	8	24	www.dsv.dk	Transport + Logistics	45,000
BG/LN Equity	11.6	56	24	www.rb.com	Household + Health	35,000
MC FP Equity	7.9	92	24	www.lvmh.fr	Luxury Goods	125,000
Brewer Equity*	15.5	46	24	anonymous	Beverages	56,000
DHR US Equity	16.1	52	22	www.danaher.com	Industrial + Medical	62,000
1363Z NA Equity	11.0	n.a.	n.a.	www.frieslandcampina.com	Dairy	22,000
ASSAB SS Equity	7.4	20	24	www.assaabloy.com	Lock/Entrance	47,000

*Price/earnings (P/E) is in the range of comparable companies within the same industry.

My 18-month career break was anything but planned. In 2014, I lost both my father and my brother. I loved them both so much, and we were very close and spoke together almost every day. When they were both diagnosed with terminal cancer, I made the decision to leave my job and spend as much time as possible with them as they took their last steps on earth. I will never regret this decision.

It seems like an impossible task to move on from this kind of grief, but at least I had some peace of mind knowing that I did everything I could for them.

During this period, I spent a lot of time at the hospice and got to know the staff very well. On long nights, I would ask them what different people talked about in their last days. The nurses told me that their patients talk about everything, with one exception. Work. In that split second, at about 2 a.m. on a December night, my whole life changed. I couldn't possibly go back to my day job in M&A when most peoples perception was that there was only at most a 50% guarantee of success. Where's the pride in that? How is that helping anyone to achieve their life's goals?

That's when Ian pointed out the madness of doing corporate M&A when the outcome is like flipping a coin (the global M&A average outcome is actually worse than just flipping a coin but never mind). He told me to put my theory into practice and when it worked, we decided to write a book about it.

The solution is business model-driven M&A, and we are excited to share this formula with you. Never again should any person consider corporate M&A to come with a 50% or more risk of failure, when we all know that a little homework can easily (and immediately) move that failure rate down to 30–40%. A little more training will take it way below 10% with repeated success, just as our M&A Elite has proven.

Ian came up with an arc for the narrative, and from there I was able to write the book from start to finish. Our complementary strengths made our partnership much greater than 1+1, and new synergies were released, as they are in the best and most successful mergers. We also owe so much to the journalist Kathryn Gaw, who has been proofreading every single word and line.

We are grateful that so many highly talented practitioners and academics joined our M&A panel:

- Cees 't Hart, former CEO of FrieslandCampina
- Jan Barsballe of ASSA ABLOY
- Marc Koster of Heineken
- Kurt Larsen of DSV
- John Schultz of Verizon
- Klaas Springer, treasurer of FrieslandCampina
- Professor Florian Bauer
- Professor Svante Schriber
- Professor Kurt Matzler
- Professor David R. King

We chose our M&A Elite after interviewing a number of global professional investors and asking them if they knew of any company which has seen continuous M&A success over a 10-year period. We used one metric to measure their success—total return analysis (TRA). We ended up with seven convincing companies, randomly chosen across countries and industries.

We then ran their corporate acquisitions or mergers through the Business Model Canvas. This was a deliberate choice to avoid any kind of corporate strategy 'blah blah,' and to get straight to the point.

It was surprisingly easy to have a conversation about M&A when it was supported by the Business Model Canvas. What also came as a bit of a surprise was how few building blocks were actually M&A drivers for the various deals, and how few metrics were actually used to decide on buy/do not buy. Simplicity and minimal metrics seem to be just two of the secrets behind successful M&A.

We were happy about our conclusions and research, but we wanted to find more data on SMEs. Through Wiley's Online Library we found some leading academics with similar interests in explaining M&A success.

As M&A practitioners, we love market values based on long-term TRA, as it is pretty hard to argue against conclusions based on this framework.

The professors also had a great belief in the fact that their global research on SMEs was hard to argue against. Contrary to the 'Financial Economic' school of thought (which is used by us in this book based on TRA), the professors had to rely on data collection from SMEs based on a 16-question SME survey. Some of the concerns you may have in such questionnaires would perhaps be: Did they include enough transactions, and were they all relevant? Was the response rate high enough? What about the non-response rate? Were there any epic failures in the world of M&A that haven't been accounted for?

We urge the reader to find this evidence for themselves. The world's leading academics did, and still do, a great job in finding ways to M&A success.

We then thought, what if we asked the Global M&A Elite to take the SME survey and compare their results with the SME responses? From a statistical point of view, it is not advisable to draw any conclusions from a survey conducted on a limited (seven) number of corporates. Still, we thought it would be interesting to take a look at the conceptual framework, and it turned out that for both SMEs and M&A-savvy global corporates there is probably a higher chance of M&A success if you have a degree of high complementarity in your business model drivers.

Well, what is the difference between SMEs and large multinational organizations as described in this book? Compared with SMEs, larger organizations have tighter hierarchies, employ explicit coordination mechanisms, offer less freedom in decision-making, and require that

M&A activities be professionalized (e.g. with large M&A departments and consultants). Nonetheless, M&A is also a strategic tool applicable to SMEs and in Europe in particular, SMEs contribute significantly to the M&A market volume. Despite the fact that SMEs usually do not have an M&A department and highly professionalized acquisition processes, top SME acquirers have a clear strategic idea on how to create value with acquisitions, similar to the Global M&A Elite.

Another interesting observation is that the global failure rate for both SMEs and large corporates seems to be 50% or more, despite the fact that the large corporates, more often than their smaller SME peers, hire the world's top-ranked M&A investment banks as advisors. Yet more proof that the road to M&A success has to come from within your own company. You cannot buy success from the outside.

What happened during meetings and calls was that our teams started to develop a highly complementary cooperation. It was great to both challenge and support each other as M&A case studies met M&A research studies. You will see this for yourself later in this book.

The Professors, the Case Insights, and the Research Insights

The following esteemed professors have all produced research which has shown that corporate M&A creates value, as well as identifying the pitfalls and probable limitations:

   Dr. Florian Bauer, Professor at Leopold-Franzens University Innsbruck

   Dr. Svante Schriber, Associate Professor at University of Stockholm

   Dr. David R. King, Associate Professor at Iowa University

   Dr. Kurt Matzler, Professor at University Bozen, Italy

Each Case Insight has been based on an interview with global equity investors and a deep dive into each company with regards to equity research, personal interviews, and a rigid TRA over 10 years, during a period in which the company has either conducted several M&A transactions or at least one game-changing merger. In each case study, we reveal the M&A drivers in the Business Model Canvas, as we have learned that the companies are all business model driven in their quest for M&A success. They choose only a few building blocks in the canvas as M&A drivers, which makes the companies able to drive (really) hard on the right issues. They choose only a few metrics in both selecting M&A targets as well as following up on deals.

Each Research Insight is based on global academic research from the leading professors within the field of corporate M&A. As with any solution in life, there are always going to be some drawbacks and the professors have been asked to challenge our conclusions (or support them) in any way possible. As most academics would start answering a question with "… it depends," the Research Insights contain a point, a counterpoint, and a contingency section. This is important, as what drives performance is not necessarily the opposite of what causes failure, and the aim of the Research Insights is to broaden the view. All Research Insights are followed by references to useful articles, books, or concepts in order for the reader to build their own business model-driven M&A formula for success.

Notes

1. Estimated failure rates are typically between 60% and 80% [Homburg, C., & Bucerius, M. (2006) Is speed of integration really a success factor of mergers and acquisitions? *Strategic Management Journal*, 27(4): 347–367] and 70–90% that do not create any value [Christensen, C.M., Alton, R., Rising, C., & Waldeck, A. (2011) The new M&A playbook. *Harvard Business Review*, March: 48–57].
2. Ranked on 3y Total Return prior to 2013. Also ranked No. 1 in Europe on 5y Total Return 2009–2014. http://citywireusa.com/news/euro-bond-s-best-of-the-best-most-consistent-managers-revealed/a785404

Part I
The M&A Formula

1 What Is the M&A Formula?

Everyone thinks that at least 50% of M&A deals end in failure. But nobody really seems to care, as long as they believe that everyone else is working with the same odds.

This is not OK. Every M&A failure costs time and money, but more importantly, by chasing bad deals you are standing in the way of your own company's value-adding growth. Done properly, M&A can be a successful way to efficiently grow a business. Done poorly, it will destroy shareholder value while making your external advisors rich, and at the very least it will make you look foolish, and may even harm your career opportunities.

This at least 50% failure rule has allowed investment banks and M&A advisors to encourage any kind of deal—whether it is appropriate or not—for decades. After all, they get paid either way, so what do they care if the deal goes bust?

This is particularly true when you agree to pay external advisors a success-based fee, as this only encourages them to attempt any deal whatsoever, thereby continuing the high global M&A failure rate. Meanwhile, CEOs, CFOs, and small-business owners are left unable to reach their true potential. If we are going to reduce the M&A failure rate, then someone has to stop doing deals. And the deals they have to stop doing are the ones that will end in failure.

The truth is that M&A doesn't have to be this risky. The best M&A dealmakers in the world have achieved M&A success by following three rules:

- Every single deal has to be business model driven.
- They involve the whole organization through great leadership.
- They do not rely on external advisors for their own M&A success.

It doesn't matter whether you are a corporate executive, an SME owner, a new employee, or a student, if you can understand business model-driven M&A, then you are on the path to success.

The late Robert Merton coined the phrase 'self-fulfilling prophecy' way back in 1948, and it is still used today. If you go into a deal believing in the at least 50% rule, then you are already accepting the possibility of failure. It will be obvious in everything you do—your attitude towards the deal, your attention to detail, and your preparation. But if you believe in M&A success right from the very beginning, then you will radiate the confidence that you need to inspire your team and drive the deal forward. This is strong leadership, but it all starts with one transformative decision: you have to decide that you are going to be an M&A success. Make that decision and this book will give you all the tools you need to make the right deals, avoid the wrong deals, and make you look like a deal-making genius. But it all starts with you.

Before the M&A Formula…

If you can sum up the purpose of your business in one sentence, then you have already taken your first steps toward M&A success. Before you start to look at the M&A Formula itself, you have to know what your company does—in other words, what are the key drivers of your business?

Maybe you are driven by creating the cheapest possible product in the market? Or maybe your key driver is to provide an increasing range of customer services?

As soon as you have identified the one or two key drivers of your business, you will be ready to start growing your business.

Information is coming online so fast, and new sectors are popping up everywhere. Some of the biggest companies in the world (Amazon, Facebook, Google, Netflix, Tesla) were hardly known some 10 to 15 years ago, and who knows what new sector will dominate the global marketplace 10 years from now.

Now, you're probably thinking, Apple does very little M&A; didn't they do pretty OK?

Absolutely: Apple's total return to shareholders in our research period is a whopping 940.8%, which dwarfs the total return of the M&A Elite.

There are two ways to grow a business: through organic growth, and through M&A activity. Build or buy. Apple is the ultimate example of an organic growth success story, as it has used its own innovation to create the sort of technology hardware which is craved by the world.

But Apple is unusual in its success. So we started to look for other firms with a high total return to shareholders over the same 10-year research period to challenge our M&A Elite. We got inspired by a publication from Heidrick & Struggles called *Accelerating performance*.[1] These so-called 'super-accelerators' are actually a fascinating mix of 'build' and 'buy' corporates (plus a couple who do both), which have increased their TRA dramatically over 10 years. The report was probably not intended as an M&A study, but we turned it into one for two reasons. Firstly, and most importantly, we wanted to compare the world's most successful firms to our M&A elite. Was build better than buy? (No) Secondly, did any of these firms actually rely on corporate M&A to become successful? (Yes)

TRA on super-accelerator firms:

- Apple 940.8%
- Google (Alphabet shares) 243.9%
- Comcast* 183.3%
- Softbank Group 252.12%
- Cigna 206.7%
- Gilead Sciences 358.3%
- Starbucks 242.2%
- Danaher* 191.8%
- VISA 526.8%
- Biogen* 476.5%
- Shire* 360.7%
- HDFC Bank* 504.5%
- Intercontinental Exchange 171.5%
- Illumina 551.4%
- Cerner* 316.4%

The Global M&A Elite has delivered TRA varying in the range of about 200–300% (with the exception of ASSA ABLOY at 327.9%). These TRAs are partly driven by M&A transactions, whereas a TRA from the likes of Google (at 243.9%) is the result of a rather genius search machine. The TRA can be expressed in another way—if you had invested $100 in Google 1st January 2007 you would have your TRA+initial investment 1st January 2017 $343.9 (100+243.9)—a similar investment in ASSA ABLOY would have grown to Skr 427.90 if Skr 100 had been invested 1 Jan 2007. What is perhaps even more impressive—the Global M&A Elite has delivered almost the same value $(331.4)^2$ to an investor as Google (343.9) in the 10 year research period.

Should You Build or Buy?

If you want to grow your business, you have to either 'build or buy.' This is the big conundrum for any CEO, big or small. Do you always have the time (and the resources) to build, and will your company always be more efficient for building as opposed to buying? For most people, the answer is no. That's why M&A is so popular at every level. No one has the time or knowledge to build themselves, so a lot of companies are buying, and it's only going to accelerate.

The highest identified TRA we measured over this period of 10 years belongs to a 'build' business model. Still, how many Apples are there in the world? Do we ever hear about what happened to all the companies who didn't succeed like Apple? Do you remember Nokia, Kodak, Xerox? Together with high-fashion clothing, technology firms and biotech represent some of the highest-risk business models, often with a binary outcome: live or die.

We have no long-term measure for all companies in the technology hardware sector, which is Apple's subsector, but in the past 3 years the TRA has been only 85%.

Another highly successful 'build' company is VISA, which has a similar monopoly over its sector and a TRA of 526.8%. But over the past 10 years the subsector 'data processing and outsourcing' has returned 'just' 207.7%.

Many of the high performers in the survey are pharma and biotech firms, and they have really made their stockholders happy; but the total pharma+bio index has only returned 130% to shareholders.

There is no evidence that 'build' is better than 'buy' or vice versa. Still, the jury is always out on corporate M&A, where a 50% or more failure rate prevails. Or so they say. Why don't you go and count the 'build' failure rate in biotech, technology firms, or high-fashion clothing? 'Build' doesn't promise better risk-adjusted returns than 'buy.' Furthermore, 'build' does not mean you are home safe, even with the best possible business model. Just as many failures prevail in 'build' as in 'buy'; you just don't hear about them so often.

Moreover, why would you ever limit yourself? M&A is about creating options, and so is running a business. Why not do both? Take a look at the companies in the super-accelerator group above marked *. These companies have seen huge success from using both the 'build' and 'buy' models. Besides, all of our Global M&A Elite companies are no strangers to creating shareholder value by 'building' up the business. Essentially what they do is just invite others to the party through business model-driven M&A.

The opportunity is huge for people and companies who can move fast and show that they are great corporate acquirers. Perhaps you realize right now that you should have started building online channels 4 or 5 years ago, and you think you've missed the moment. Think again. Corporate M&A allows you to travel in time. You can partner up or buy people and companies with the competencies that you need in your existing business model. It is that easy.

Deciding to Buy: Growing Your Business with M&A

This book won't tell you how to 'build,' but it will make you a successful corporate acquirer. You can avoid M&A failures with immediate effect and gradually build M&A competencies into your business model, moving from a defensive stage to an offensive stage, and elevating your competencies to the level of the Global M&A Elite.

While you were setting out your business model drivers, you will have set a number of stretch goals for your company. Now is the time to revisit them.

With these stretch goals in mind, you will be better able to create your M&A target list, and begin the crucial process of running these targets through the Goldman Gates.

Don't be alarmed if your long list of target companies is whittled down to just one or two by the end of this process. That's kind of the whole point. Stopping your company from doing the wrong deal is the first step towards reducing your M&A failure rate. When you stop doing the wrong deals, your failure rate will become much lower and your company will become 'anti-fragile,' as you learn from your anti-portfolio.

Every time you run a target through the Goldman Gates, you will be able to allocate an M&A score, which tells you whether the target is a 'go,' a 'no go,' or a 'maybe.'

'Go' Targets
These are the firms which are going to be the best fit for your company. Move forward now before someone else beats you to it!

'Maybe' Targets
You need to do more research into these companies, so you can identify the blockages and work out whether or not they will be a good fit. Don't move forward unless you are satisfied that your due diligence efforts can promote these companies to 'go' targets.

'No Go' Targets
These are the firms that you should avoid at all costs. However, this doesn't mean that you have wasted your time. These names will form the basis of your anti-portfolio—a vital document which will keep you and your colleagues from making the same mistake in the future. Saying no is not as easy as saying yes. However, you must learn to say no a lot more because that's the only way to avoid doing the wrong deals. Besides, being absolutely clear about what M&A deals you don't want to pursue only sharpens your own M&A formula, and makes your team much more directional and decisive on future M&A deals.

Once you have identified the 'go' and 'no go' deals, you move them to the M&A Launchpad. 'Go' deals get onto your Launchpad, and 'no go' deals are added to your anti-portfolio. Think of this as a big bin beneath the rocket-launch platform—occasionally you can take a look at it and remind yourself why it never became airborne.

Look at your company's stretch goals and business model drivers, and create a version of the M&A Launchpad which is unique to your company. Then all you have to do is follow the rules. Follow the M&A Formula and run each deal through the Goldman Gates.

The Three Steps

The M&A Formula is a simple, three-point plan that will ensure you are in the best possible position for M&A success before you even know what your next deal will be.

Whether you are the head of a multinational industry behemoth, or a small-business owner mulling your first acquisition, the M&A Formula will keep you and your team on track and immediately reduce your risk of failure.

> ### *The Three-Step M&A Formula*
>
> 1. Follow business model-driven M&A.
> 2. Exercise strong leadership and communication.
> 3. Take ownership.

Follow Business Model-Driven M&A. Focus on a few key, high-impact drivers for your business. They must be long-term drivers over at least 2 or 3 years, and preferably over a longer period of 5 to 10 years. This isn't something you have to do every time you begin a new deal; it should be part of your long-term corporate strategy.

Every company will approach this in a different way. For DSV, the main driver is cost, and due to the company's reputation for operational excellence, it is well placed to achieve this goal. When they buy other firms, it is because they aren't as operationally good as DSV, so there is an opportunity to save money by bringing them into the fold.

Your company's main driver might be completely different; it all depends on the results of a deep dive into your business model. What building blocks can you turn into M&A drivers? Choose only one—or at most two—as you want to stay focused and drive hard on a few metrics rather than spreading yourself too thin. Any building block can do it, as you will see later. No specific country or region applies. It's not about branded products or intellectual property rights either. It's about choosing the right business model driver for your company, getting your own organization behind it, and making sure that you secure your long-term M&A funds for the projects.

Exercise Strong Leadership and Communication. Communicative leadership in corporate M&A doesn't happen overnight. It is a skill which you have to work on, and embed in your company. Still, the formula itself will kick-start clear communication with its no-bullshit focus on business model drivers. It is about creating a winning mindset in your organization through a sense of belonging. Every person in the company should be able to say: "I know why we do M&A and I know what is expected of me."

Take Ownership. No external advisor will ever make you a successful corporate acquirer.

The reason that you have to take ownership is not because you're better than anyone else in all M&A disciplines, it is because no one else will know your business model as well as you. You are at the front line. Your own organization knows your business better than anyone.

Understanding Business Model-Driven M&A

The M&A Formula (see blogs on this subject on www.fixcorp.co + other references) requires some basic knowledge about your firm's as-is business model.

The Business Model Canvas is a useful tool for this conversation, as it is basically a corporate diagnostic drill which must be based on the common knowledge of your firm (see Figure 1.1).

Never underestimate the negative consequences that you could experience if your team is not aligned in the mutual understanding of your business model.

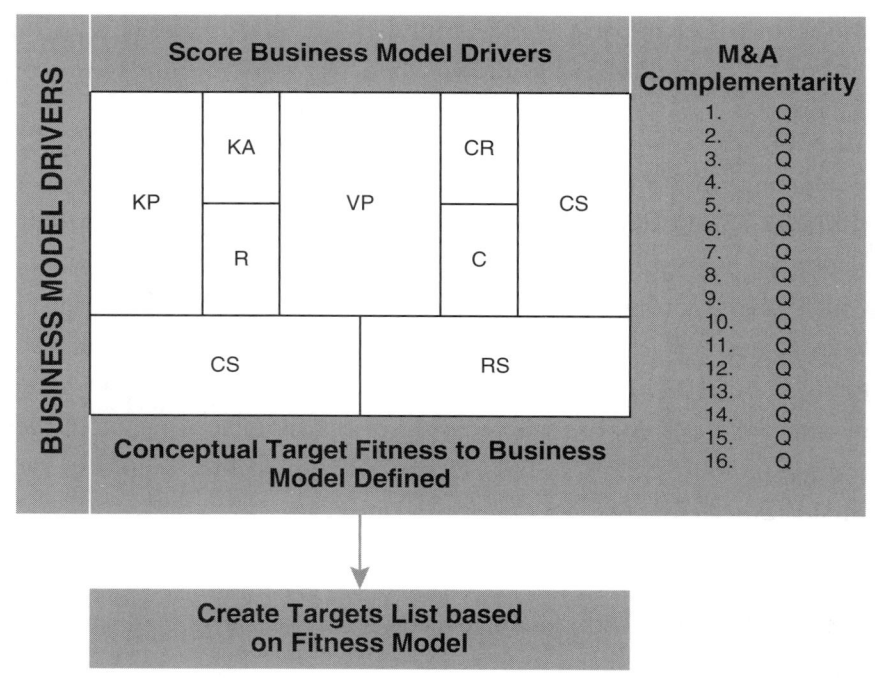

Figure 1.1 Business model drivers

CS: customers; CR: customer relationships; C: channels; VP: value proposition; KR: key resources; KA: key activities; KP: key partners; CS: cost; RS: revenue

Applying the M&A Formula

#1 The business model and the M&A drivers (Business Model-Driven M&A)

To quote Lewis Carroll's *Alice in Wonderland*: "If you don't know where you are going any road can take you there," and this is particularly true in business model-driven M&A. Thanks Erik (Elgersma).

We encourage you to discuss your business model drivers *before* moving on to the M&A Formula itself, where the first step is to choose one or two M&A drivers in your business model on which you can leverage and create value with M&A.

Case Insight

The global car firm, the 'missing alignment,' and how a Business Model Canvas discussion got the team back on track…

The top 50 management team of Toyota Financial Services (TFS) was gathered in one huge meeting room with their advisor Business Models Inc. (CEO Patrick van der Pijl).

Read the full story on: https://www.businessmodelsinc.com/client-story/toyota/

Julia Wada, group vice president of HR and technology recalls how many different opinions were out there on fundamental issues such as "who is the customer?"

On the one hand you can argue, that any car OEM (Original Equipment Manufacturer) must consider the person behind the wheel, "the customer" (a person driving a Toyota).

(continued)

(*continued*)

On the other hand you can argue, that any OEM supporting car-finance (to increase their car sales) must consider the OEM re-seller network as "the customer" (Toyota dealer selling a car to a person or a company perhaps).

Finally the former CEO of TFS, George Borst said: "We can agree to disagree but I believe these are both our customer."

The most interesting point from this book's perspective is really, that the M&A Formula will not work if your own organization is not aligned on your own business model. Adding a new business to an existing, but blurred business model really compares to building a house on a shaky or muddy foundation. Make sure your groundwork is solid before using the M&A Formula. More often than not companies, or more precisely their organization, are not ready to do Corporate M&A simply because of this single issue. No post integration plan, due diligence whatever any M&A sellside will sell from their tool box will ever fix this problem. Never.

Most discussions based on the Business Model Canvas actually start with "who is our customer". And almost as many times, according to Mr. Pijl, there are actually different views on something so fundamental as who is our client really?

Kindly remember there are a total of nine building blocks in the Business Model Canvas presented in the book Business Model Generation, produced by Mr. Pijl.

Our point is really this. Make sure your whole organization is fully (and truly) aligned with all building blocks before you start using the M&A Formula itself. Not only the customer discussion which is crucial in itself though.

This is a real life experience and it is definitely not an unusual occurrence at all.

> Firstly, go rigidly through all building blocks with your team, be it 2 people or top-250 and make sure you discuss all options.
> Secondly, the Business Model Canvas is a highly useful tool for this discussion as all participants can easily engage in the discussion that becomes very structured.
> Thirdly, and perhaps the best result, with this approach you avoid personal beliefs or people having a special position in your company (read later "what you don't know about biases will harm you").

If you do not feel confident enough, then you should consider hiring an external advisor who can take you and your team through this discussion before you move on to the M&A Formula itself. I have seen Business Models Inc. in action, and I can really recommend their methodology based on the Business Model Canvas itself.[3] As a team, you need to have a common understanding of where you add value in your business model. If not, you will have a hard time narrowing it down to one or two business model drivers in the first step of the M&A Formula.

Say you are running a gadget firm, where your greatest strength lies in making items at a very low unit cost. You discuss your company goals and try to narrow your business down to one or two building blocks, which you will leverage on in terms of M&A. In this case there is only one building block: *cost*.

Now comes the question: "Do we have to be the leading gadget-making firm in the world to become successful in M&A?" No. You just have to be better than your closest competitor.

Imagine two people on an outback trip in Canada. A very hungry grizzly bear suddenly approaches. One guy quickly straps his running shoes on, but the other one tells him "you idiot, no human being can possibly outrun a grizzly bear! You don't have a chance." The first guy replies: "Actually, I only have to run faster than you …"

If you run faster, if your firm is better than the company down the road, then you have a chance of creating value in M&A. You don't have to be the world's best in your category, as long as you are better than the company you are planning to buy. Just make sure you know what business model driver you want to apply.

Leadership and Role Models. Once you have chosen one or two business model drivers, it is time to find a role model, perhaps from the Global M&A Elite or an SME firm with proven M&A success. Discuss with your team what your role model is doing. Could you do the same, and why? We have provided you with seven possible role models from the Global M&A Elite, but there are several hundred firms in the world who have seen repeated success in M&A. The best role model for low unit costs is probably DSV, although the company would probably describe its strength as 'operational excellence,' as it always gives its clients an excellent service in bringing stuff from A to B. This company started out with ten trucks and has grown to almost €10 billion in revenues by making acquisition after acquisition whilst delivering, like the rest of M&A Elite, a high total return to shareholders.

M&A is about doing; it is not like building a firm. Buying growth must be clearly articulated in terms of why you do M&A, what you expect, when, and why. Without this clear mission statement, you cannot lead an organization in M&A and you will not be able to follow up on your transactions.

Taking Ownership. Your team is aligned on your business model and you have chosen one or two business model drivers in M&A. You have chosen a role model from the M&A Elite companies, perhaps a combination, and you have also chosen one or two key metrics to align with your shareholders. Now, you just have to apply the rest of the ingredients and you have baked-in the M&A Formula for future successful growth.

The Goldman Gates will help you validate various targets and see if there is a business model fit. Once you have identified the M&A complementarities, you can go directly to the M&A Launchpad (more on this later).

If you are an SME business owner, you are the boss and you have majority shareholdings, you can start right away. If you are CEO, you could probably also start this transformation straight away and become business model driven.

It gets more problematic if you are a business student or a junior employee. Just remember this: the common knowledge of your organization is greater than the knowledge of any individual. The M&A Formula is based on your organization's common knowledge and it has been proven to reduce the M&A failure rate and, when organizational learning starts to kick in, you will be able to repeat M&A success. You will be right in the long run. If you seriously think, as a young and promising employee, that your company is not prepared to take its M&A activities seriously, then it may be time for you to change the company you are working for. If you spend too many years working for a company with a normal M&A failure rate, that doesn't bode well for your career.

When you buy with your heart, you are leaving yourself vulnerable to a million unknowns. When you buy with your head, it's a lot less romantic, but you are minimizing your risk of failure.

Break your decision-making into smaller pieces and thoroughly analyze each one of the building blocks. That's what we do through business model-driven M&A, and it works.

Ask yourself: is your business listening carefully enough to the people on the front line—the ones who are actually selling your products and services? Do you thoroughly consider your client segments and observe their behavior? Do you match this with your clearly defined value proposition? It is extremely important to do this on *both sides* of a potential deal before determining a fit. A good fit is the first step, but there must also be willingness, on both sides, to pursue one 'business model.' Create transparency and do not be afraid of over-communicating. There is no such thing in M&A.

Listen to the naysayers in your organization, the ones saying "why on earth would you expect that to work?" Then let them sketch other business models and create new options.

Our mission is to lower the failure rate in M&A, and we strongly recommend that you use the M&A Formula before doing anything else.

Success and Failure in M&A

There are a few different ways to 'fail' in M&A, and not all of them are bad. In fact, we would categorize failed deals as 'Really Bad,' 'Bad,' 'Bad-ish,' and 'Good.'

Failed Deals

Really Bad

Deals that were done where the outcome drastically under-delivered on expectations. Maybe the synergies were not realized, teams did not align, or governance issues killed the gains. These deals did not use the M&A Formula, instead relying on sheer monkey business!

This sort of failure is really bad; it destroys careers and loses shareholder value.

Bad

Deals that were lost because the buy-side was unprepared in some way, and the opportunity to gain was lost. There was an M&A Formula fit, but no supporting M&A Launchpad.

These failures are bad, annoying, and symptomatic of poor M&A practice.

Bad-ish

Highly desirable deals which met with the M&A Formula and were Launchpad ready, but failed because of externalities.

This usually means that you were able to expose a previously unrecognized and totally unexpected issue that would have hurt you later, so you got lucky. This is not a bad deal, it's just bad-ish.

It has to be a really big event, such as Brexit or the Swiss Central Bank removing the cap on its currency, causing a 20% immediate appreciation. We sometimes refer to these events as material adverse changes (MACs) in a deal. MACs can even be built into your M&A Launchpad, protecting your firm even during binding offer stage!

Good

Deals that were in line with your M&A Formula, and therefore had to be attempted. However, factors that would have

(continued)

> *(continued)*
> eventually turned the deal were identified early on (e.g. the seller disabling your business model drivers, culture, governance, Ts&Cs, due diligence, price, or an interloper risk). These deals are usually stopped due to Launchpad or governance issues.
> This is a good failure, as you were able to fail deliberately.

If anyone tells you that you will fail in at least 50% of your M&A deals, tell them they are dead wrong, because you are simply not going to make the same mistakes as so many others.

In this book, we have shown you that 56.1% of companies in an SME survey found that their M&A activities in the past 5 years had contributed 'significantly' to corporate success. So, we know it is actually not that hard to immediately lower your (expected) failure rate of at least 50% if you start doing the right things.

But is 56.1% so much of a difference from 50% or less? Well, yes!

If we compare this 56.1% score with fund managers in the stock-picking business, we get a good comparison. The term 'monkey business' may have been invented by the economist Burton Malkiel in his 1973 book *A random walk down Wall Street*, where he proclaimed that a blindfolded monkey throwing darts at a newspaper's financial pages would do just as well as any equity fund manager.

Many years later, the *Wall Street Journal* decided to test Malkiel's theory in real life.

It turns out that it wasn't actually that easy to use monkeys, so they used blindfolded *WSJ* personnel to throw the darts instead. After 100 tests, the result was that the professional investors won 61 times out of 100, or 61% of the time. So, the world's top fund managers are only about 10% better than flipping a coin. Really?

In M&A and the capital markets, this is actually a big difference. If you get it right 40–50% of the time, you are a monkey. But get it right 60–65% of the time, and you could be one of the world's best fund managers or an M&A rainmaker.

Success in M&A is something completely different from just avoiding failure. That's why we invented the M&A Formula. If you had invested €100 in Google 10 years ago, you would have received an annual return of about 13%, equal to the average total return of our Global M&A Elite. Google is a fantastic firm, and so are the Global Elite firms. They create value for their shareholders. They also create value for our societies in paying taxes, creating jobs, securing pensions, making great products- and services for their customers.

Notes

1. Price, C., Toye, S., Hillar, R., & Turnbull, D. (2017) *Accelerating performance*, Heidrick & Struggles, London.
2. FrieslandCampina not stocklisted + local currency the rest.
3. www.businessmodelsinc.com; their CEO was the producer of Business Model Canvas.

2 Be Business Model Driven (Step 1)

What Is Business Model-Driven M&A?

If you bake in the M&A Formula at the start of each deal, even your failures will be a success. Sometimes deals will simply fail for reasons beyond your control, and that's OK. But in most cases, the warning signs are there right from the very beginning. If you follow the rules of business model-driven M&A, you can streamline this process and transform from a defensive strategy (where you are acting to avoid failure) to a pro-active offensive strategy (where you are chasing success).

In Figure 2.1, we have spread the 16-question M&A Complementarity model in the Business Model Canvas in order for you to see how comparable they really are. The advantage of M&A Complementarity, when added to the M&A Goldman Gates, is that you can further assess if you are on a realistic mission to M&A success. You could also receive a warning sign if you are not.

Business Model Complementarity and Signals for Success

When we decided to adopt an 'M&A complementarity' score in our M&A Formula, we faced a few practical issues.

Firstly, we wanted to do a 'cross-check and report' of the two M&A frameworks ('Business Model-Driven M&A' and 'The Antecedents of

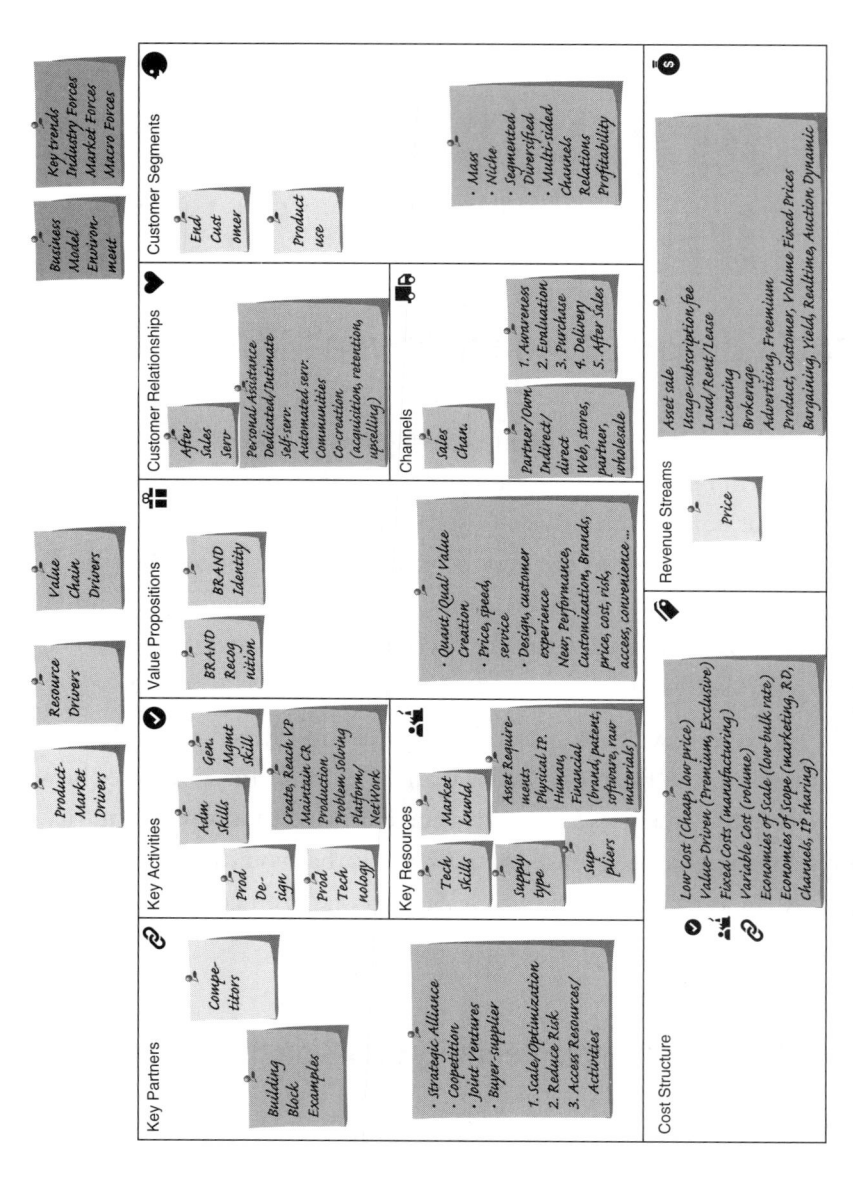

Figure 2.1 SME survey in the Business Model Canvas

M&A Success'). The solution to this was to simply ask our M&A Elite to fill out the same 16-question survey that was given to our SMEs.

The results were highly interesting, as we found that the successful, M&A-savvy Global Elite seemed to follow the same recipe for M&A success as the top-performing SMEs, and vice versa. In each case, a high complementarity spelled a higher chance of success. Danaher was the exception that proved the rule—it had a low complementarity but, as we have said before, when you look at the reasons for M&A failure, it does not mean that by doing the opposite you will see M&A success. We refer to the deep dive into Danaher's M&A activities in the Case Insight and Research Insight.

Secondly, we had to solve a few practical issues, such as the fact that the framework was also used for manufacturing firms and how much 'after-sales service' is really involved in (for instance) making an ice-cold beer or a pain killer for a headache. We decided to leave out this question for some firms, as it was perhaps more relevant for manufacturing firms than service firms or food and beverage firms.

Thirdly, and perhaps the biggest issue, what should we do with the competitors?

This became a rather controversial matter.

On the one hand, the professors on our M&A panel had listed this question as highly relevant in the M&A complementarity score.

On the other hand, the authors behind the *Business Model Generation* have several times pointed out (rigidly) that "competition is not part of your business model."

However, in the end, we decided to go with our own M&A panel's suggestion: competition is highly relevant in the M&A Formula and it is a key part of successful M&A. But it is also highly logical for a practitioner or academic with insight in both corporate 'build and buy' behavior that competition must be considered part of your business model canvas.

Inasmuch as we like the Business Model Canvas, the framework seems to totally ignore the fact that corporate M&A is a possible enabler of your business model (or if you don't follow an M&A formula for success, a potential failure). Corporate M&A makes it possible to change your own building blocks in your business model and even make some external building blocks *part of* your business model. If you rigidly follow the original outline for the Business Model Canvas, you only commit your company to build, which is a crazy idea as it suggests that no-one in the world could possibly produce any product or service better, faster, or cheaper than your own company.

TechCrunch acquired the company Here in March 2015, and the deal was finally closed in December of that year. Here produces digital mapping services, and TechCrunch rightfully feared the interloper risk from companies like Uber or even Apple. The main reason that Here had become such an interesting company was probably down to the fact that self-driving cars are already on the streets today, and development will explode in the coming years.

What is highly interesting here is that TechCrunch is owned by a consortium of German carmakers (BMW, Volkswagen, and Mercedes).

When three cut-throat rivals suddenly buy high technology which is able to collect more than 100 billion data points a month, it makes me think of a magic wand in relation to the somewhat static Business Model Canvas that ignores corporate M&A.

Firstly, the day before deal announcement, 'competition' was outside the business model of BMW, Mercedes, and Volkswagen. The day after they became partners, at least with regard to building the future of the self-driving car. The three German car producers could suddenly stand together and define their own business model, and all agree that their competitors now included the likes of Uber and Apple.

Be Business Model Driven (Step 1)

Secondly, the three carmakers suddenly got access to a technology which they didn't have in-house. In fact, it was a technology they had completely overseen, as the German engineers were probably more focused on the next Le Mans, or Formula One race.

We call this 'travelling in time' with corporate M&A. You didn't see the next technology jump, but you repaired this oversight with an M&A transaction.

The M&A complementarity score is divided into three groups, which you must score:

- Product Market attributes^{PM}
- Resource attributes^{RS}
- Value Chain attributes^{VC}

Customer Segments

Your End Customers^{PM} and the Product Use^{PM} are very important factors in both your *as-is* business model and M&A complementarity, as two companies with high complementarity will more easily understand each other when facing their new mutual client segments post-M&A deal.

Channels

Sales Channels^{VC} as a category may speak for itself. There are often obvious cost synergies in overlapping sales channels (see RB) or in getting access to new channels (see LVMH).

Customer Relationships

The M&A complementarity does not score much on these soft factors, which could be assistance in the shop, automation, or self-service, but we find that Aftersales Service^{VC} belongs in this building block.

Value Proposition

The author behind Business Model Canvas published a new book just focusing on value proposition, so this is obviously a very important factor in your business model. LVMH has truly understood how to keep this value proposition intact in their M&A activities, while RB has cleverly *improved* their value proposition. In corporate M&A, this has much to do with branding—Brand Recognition[RS] and Brand Identity[RS].

Key Activities

Companies like ASSA ABLOY make their R&D available for newly acquired firms and thereby increase profitability with better Product Technology[PM] and Product Design[PM]. The Administration Skills[RS] and General Management Skills[RS] also play an important role for many successful acquirers, but it depends on the need for management retention in the newly acquired firm. DSV would normally replace management immediately, and has a low complementarity score as a result. However, this is considered to be an opportunity rather than a threat in their M&A activities. Other firms, like ASSA ABLOY, simply acquire businesses and their management to act as business owners post-deal, and therefore have a greater need for complementarity.

Key Resources

Most companies from the Global M&A Elite have an ability to immediately lower procurement costs after a corporate acquisition, by at least 2% or 3%, because of their bargaining power and strength in the market. When there is high complementarity in the merged businesses on Suppliers[VC] and Supply Channel Types[VC], highly predictable cost synergies emerge in corporate M&A. Market Knowledge[RS] and Technical Skills[RS] also play an important role, and can add value when there is high complementarity.

It is perhaps interesting to notice that 50% of all M&A complementarities are to be found in just two business model building blocks—Key Activities and Key Resources. Larger firms, not only global firms, will often be able to squeeze out cost synergies in M&A deals in those areas, even with a short time horizon, which is the reason why it plays such an important role when diagnosing the overall M&A complementarity of two firms who may be able to do great things together (and create value for shareholders). All of these improvements feed directly into the cost structure.

Key Partners

Competitors^{PM} can become a part of your business model overnight through corporate M&A. Do not look at the world statically. In the FrieslandCampina Case Insight we saw a transformational merger in the dairy industry between two cut-throat rivals, Campina and FrieslandFoods.

Even technology can be acquired and, though it appears as an external variable in the Business Model Canvas, it too can become part of your business model with a little magic from your M&A wand. Just remember to read the instructions first (the M&A Formula).

Three Real Business Models: M&A Drivers and Complementarity Score

As previously mentioned, what drives performance is not necessarily the opposite of what drives failure. While integration for instance, has beneficial effects like the elimination of redundancies, it might also destroy value by disrupting employees, triggering negative behavior, or increasing turnover of key employees. The same holds true for complementarity. In cases of low complementarity, there might be a different deal logic, but if firms set the right measures, they might succeed. To make a long story short, statistics are helpful for understanding relationships but they are no guarantee for success, as each case can be a statistical outlier not covered by the probability of a model.

Professor Dr. Florian Bauer

Here are just three real-world examples from DSV (Figure 2.2), Danaher (Figure 2.3), and FrieslandCampina (Figure 2.4), taken from our research comparing global corporates with SME firms. As you can see, there is very high complementarity in the 16 questions, which suggests that you have an even higher chance of M&A success. Simply put, the higher your score, the greater your chances of M&A success. If you compare two possible M&A targets—one with an M&A complementarity of 3.6 and the other with a score of 4.6 (on a 1–5 point scale)—then statistically, the latter M&A will have a higher success score (1–5 point scale). In fact, just by moving the complementarity by one point, you are increasing your chances of M&A success by 7.4%.

What Are the Goldman Gates?

We have mentioned the Goldman Gates a few times already in this book, and you might be surprised to learn that they have absolutely nothing to do with Goldman Sachs—one of the world's top investment banks. Instead, the Goldman Gates are a strict M&A deal-scoring system, with its origins in medicine (more information on www.fixcorp.co).

The 'Goldman Algorithm' was examined in some detail in *Blink* by Malcolm Gladwell—a brilliant book about making decisions without actually making decisions—and this is a cornerstone of business model-driven M&A. When you give away responsibility in M&A, you can focus on the right deals and take them to the M&A Launchpad, and quickly funnel the wrong deals into your M&A anti-portfolio.

The 'Goldman Criteria,' invented by cardiologist Lee Goldman, is a widely used algorithm which helps physicians decide if a patient requires immediate hospital admission. First tested some 20 years ago, it is a rules-based tool which can considerably speed up

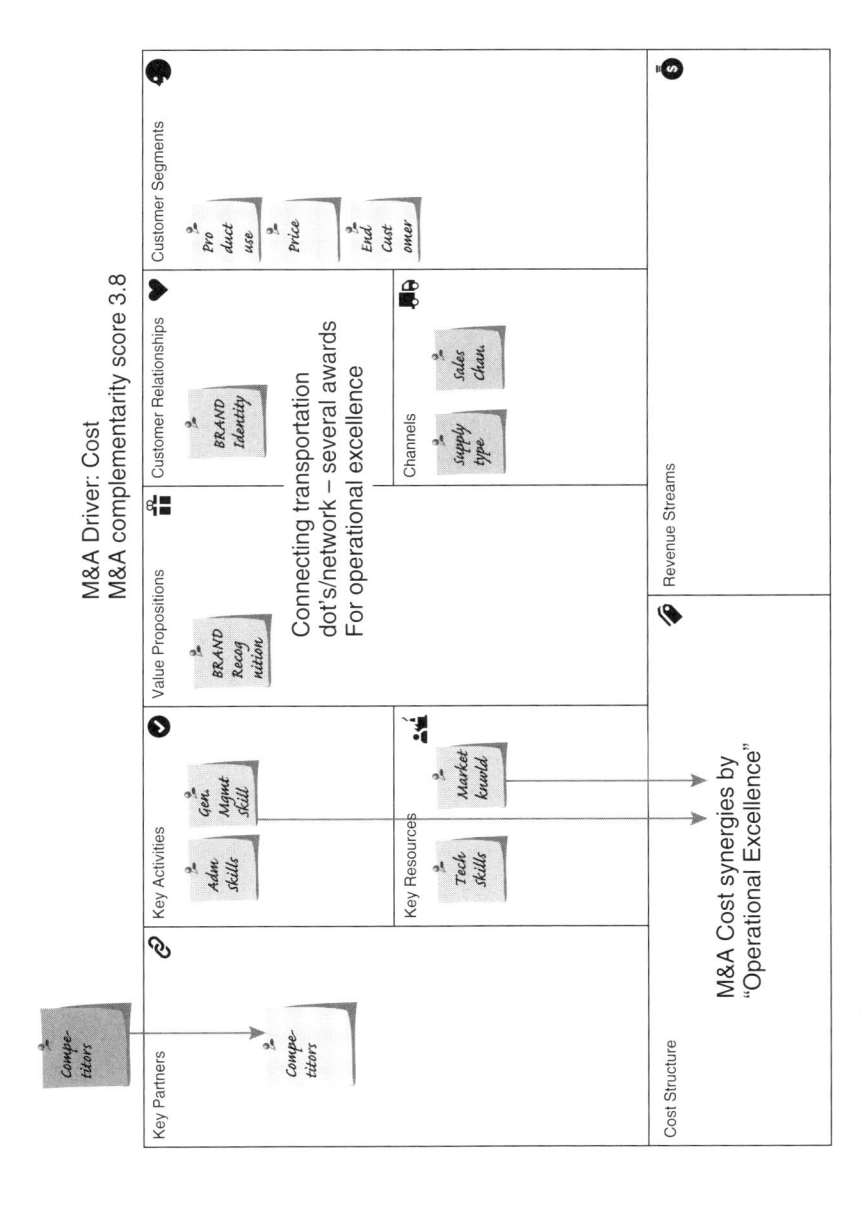

Figure 2.2 DSV and M&A

M&A Driver: Key Activity (homegrown DBS concept)
M&A complementarity score 1.5

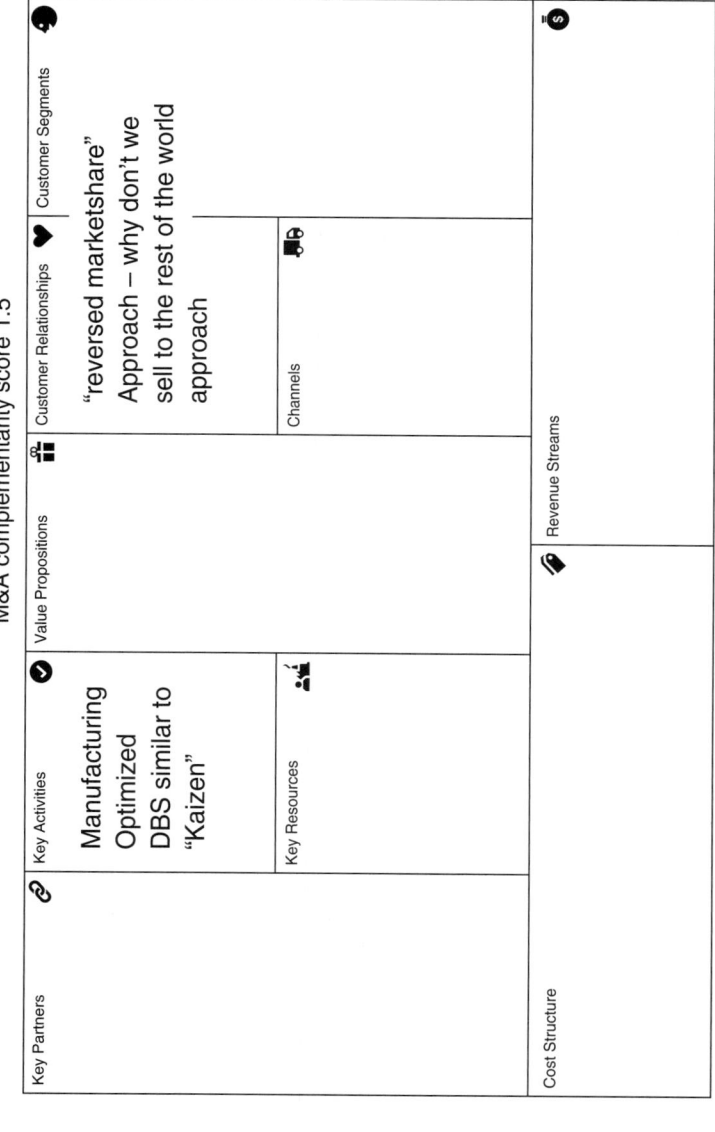

Key Partners	Key Activities	Value Propositions	Customer Relationships	Customer Segments
	Manufacturing Optimized DBS similar to "Kaizen"		"reversed marketshare" Approach – why don't we sell to the rest of the world approach	
	Key Resources		Channels	
Cost Structure			Revenue Streams	

Figure 2.3 Danaher and M&A

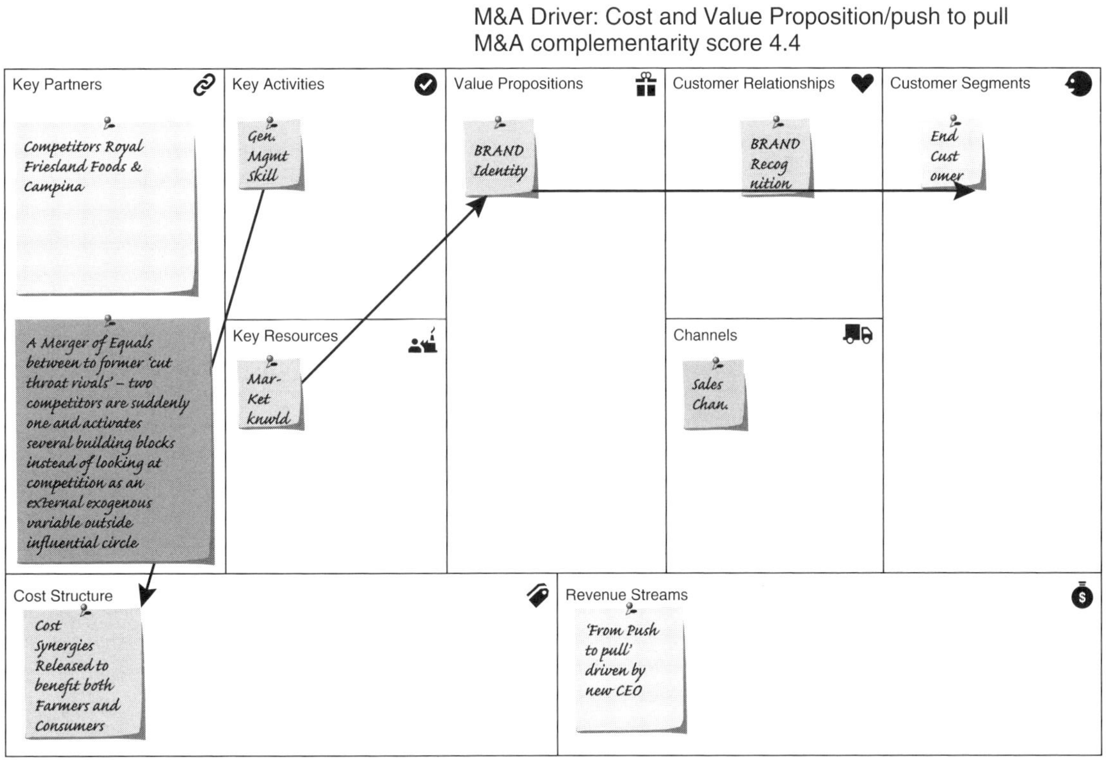

Figure 2.4 FrieslandCampina and M&A

61

the decision-making process, particularly when it comes to heart conditions.

In the Introduction, we discussed how the world's best money managers are also rules-based. They follow a strict investment process, which ultimately leads to a 'go' or 'no go' choice. Being 'just' 61% right makes a huge difference in terms of return for professional investors in capital markets, and could mean moving up in the field from top 1000 to top 100, and so forth. You need to make these judgment calls and, when they are rules-based, you will make more right choices than wrong choices, so long as you have correctly defined your first step in the M&A Formula, which is business model-driven M&A.

The Origin of the Goldman Gates

First, the physician establishes the existence of four key facts:

1. Is the ECG result abnormal?
2. Is there evidence of unstable angina pain?
3. Is there fluid in the lungs?
4. Is the systolic blood pressure under 100?

You don't have to be a physician to know that if the answer is 'yes' to two or more of these questions, then the patient will need to be kept under close observation.

This decision is made without a massive stream of data, or an expensive analyst. It is a simple way to quickly assess whether or not a patient needs extra attention.

The Goldman Gates approach is about getting your M&A projects to 'go' or 'no go' instantaneously. This process is much simpler than most stuff I have ever seen in corporate M&A, especially when firms call it 'strategy,' but it serves the same purpose as the original Goldman

Algorithm—spend time on the deals you want to do and stop wasting time on stuff you don't want to do.

What I see in many corporate organizations is that some of them spend too much time on deals that will probably never materialize. This is mostly caused by the fact that the foundation for doing M&A has not been agreed on from the top down. In the event that the top management is actually aligned, which is rarely the case in reality, you still need to include level-2 people and other employees who are further down the chain. This is just as critical as having both surgeons and nurses in a hospital, where everyone needs to understand the end goal and the process in order to save lives, or become a successful corporate acquirer.

The Goldman Gates are there to protect your business from the wrong deals and allocate as many resources as possible to the ones that have a business model fit. By now, you will have identified your own business model drivers, and you will know from the M&A Formula that you are ready to start making deals (see Figure 2.5).

How to Create a Target List Based on the Goldman Gates?

How you score your individual business model drivers is not really that important. Some companies use low–medium–high, 1–5, or any other scoring system. What is important though is that you start building your own internal M&A scoring system based on business model-driven M&A.

That sentence alone killed two birds (maybe even three) with one stone in relation to the M&A Formula. Firstly, you force yourself to implement no-bullshit frameworks, as business model-driven M&A is always much more granular than corporate 'blah blah.'

Secondly, you are taking your first step towards DIY M&A, which is crucially important for M&A success as no external will ever do this for you. The whole drill of using our M&A Goldman Gates is to create

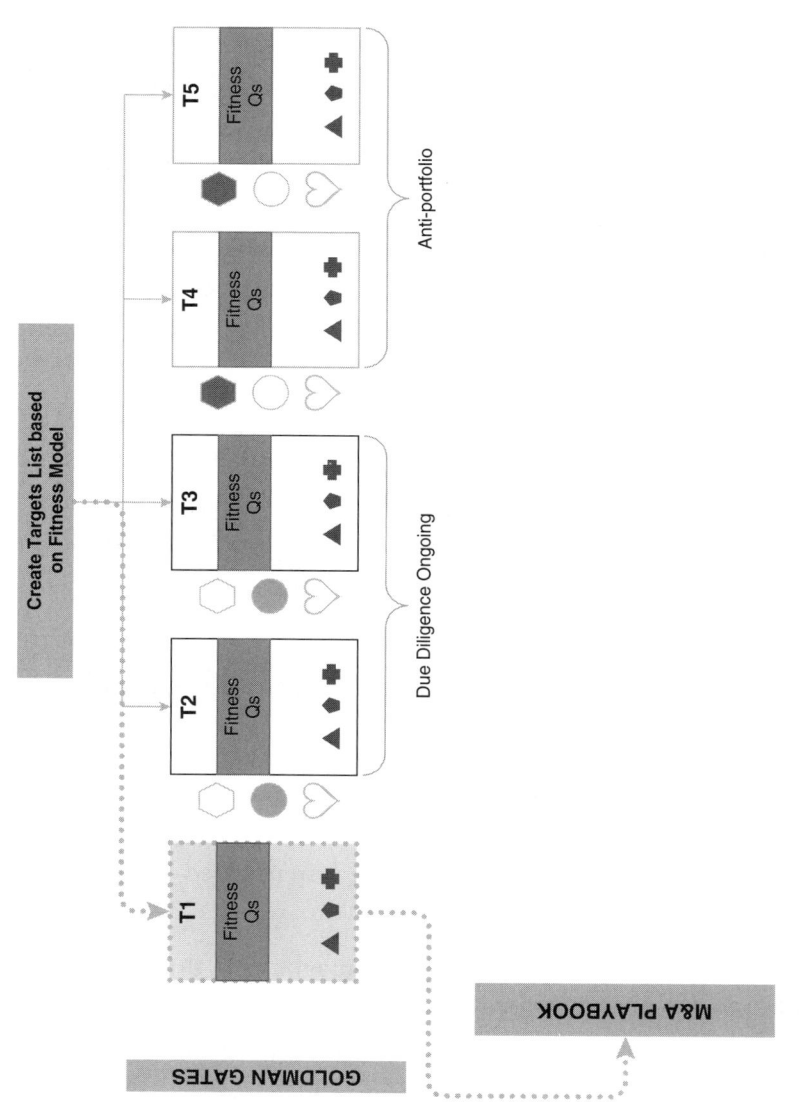

Figure 2.5 Goldman Gates Scoring model

a network-based organization, which is the human side of M&A and the last pillar in the M&A Formula.

Another member of our Global M&A Elite (the global brewer) is a good example of higher granularity in M&A Goldman Gates. Through our interviews, we learned that this company drives hard on four M&A archetypes. Each one of these deals is rooted in different business model drivers, but they also have different time horizons and quite different Goldman Gates criteria beneath the business model driver. The global brewer is in fact a perfect example, as some of the minimal requirements like global market share (cost synergy deals) are not representative for other kinds of M&A deals (like access to a new market).

One of the key advantages of having different archetypes of M&A deals with a different business model approach is that they are also comparable with historic deals done. Some M&A deals are more similar than others, and this makes you more comfortable as you can get a better sense of whether or not you are on a realistic mission.

If your foundation for business model-driven M&A is strong enough and baked into your target selection process, you will not only do the right M&A deals, you will also spend much less time on the deals you should never do.

With the global brewer as an example, you could list 'Market Share' as a business model fit, and based on this you might score the expected cost synergies within 2 years as 'highly likely' or 'less likely,' and develop a rules-based scoring system from there. It works for the global brewer, maybe it could work for you too!

Many companies calculate their financial scoring based on the size of the company; however, it can be much more enlightening to simply look at revenues, so that you know just how many 'days of revenue' you are buying. Does this deal move the needle? If it doesn't move the needle, ask yourself if you are going to make a series of acquisitions similar to this transaction, which is also known as a roll-up strategy.

If not, why bother in the first place? If you do corporate M&A, you want to make a positive, noticeable impact, otherwise go do something else. An expected long-term impact on earnings per share may also come in handy as a way of illustrating how this deal helps to 'move the needle' for your firm.

Of course, every company will have to consider elements such as geography, product areas, and key financials, but you will add your own driving factors according to your own business model. We find the best approach is to simply follow your existing business model canvas. Not only does this provide much more granularity than corporate 'blah blah,' it also has the luxury of visualizing the individual business model fit. Some people prefer visualization to an Excel spreadsheet, with synergies in sheet 17, row 73, below EBIT and working capital, and so on … you get the picture.

As soon as a new M&A deal has been properly categorized, you can add in any extra information and compare this intelligence with previously done deals, both internally and externally. It's also worth comparing price levels with so-called trading and transaction multiples to get an idea of the potential premium that must be paid. This will usually hover somewhere around 20–30% above the trading price. However, we have seen deals where companies have been taken over basically for free by just agreeing to take over the total debt (minus a haircut of 30–40% on bank debt!). Despite the 'bargain price,' these M&A deals were failures. You might think you are getting a bargain, but unless your business model driver is to buy cheap firms and transform them, this is rarely a good idea.

Once, at FrieslandCampina, the company paid almost 100% over the stock-listed price and both the company and the seller are still extremely happy about the deal today. Success! Forget about the price—it's not important. It's also not about valuation, these are just hygiene factors. By all means get them right, but this is just a small element of the process. Business model-driven M&A is all about getting your corporate foundation right. Too many people spend too much time on articulate knowledge or the small things in M&A deals. We all read the same books and went to similar schools. Being a master in valuation gets you absolutely nowhere. It's just stuff you have to know.

Every successful corporate acquirer has a different way of scoring their M&A deals, but the organizations will thoroughly discuss 'go'/'no go' based on their business model drivers and supported by the Goldman Gates.

No Go
No business model fit (or at the later stage due diligence finds a number of major issues with the deal and the decision is made not to proceed). This company is added to your anti-portfolio of firms that you will not consider again.

Hold
More knowledge of some kind is required before conducting Goldman Gates.

Go
Proceed with this target to the M&A Launchpad. Goldman Gate-approved (business model fit, individual criteria, and M&A complementarity).

Now you have your M&A Launchpad (more examples on www.fixcorp.co), but you also have your anti-portfolio, which is just as valuable. So how do you learn from an M&A anti-portfolio? Let us provide an example. When you write a book like this, you will find that 30–40% of the stuff you write does not really fit in, or may have a wrong fit to the book. Sometimes the company you thought was great turns out to be not so great. Other things, such as related research, turn out to be outdated or have flawed data. All the stuff that doesn't have an immediate fit with the arch of the book was placed in our so-called 'parking space,' which is similar to the M&A anti-portfolio. In the making of this book, we started to build a stronger focus as we now and then looked at the parking space and learned by it. Oh yes, stuff like that doesn't fit. We grew stronger and became 'anti-fragile.' We have introduced a global myth killer, an M&A Formula for success. Of course, the jury will be out and challenge us. In a company, you will also learn from your M&A anti-portfolio.

"I remember two companies in my previous job that were just about the best business model fit I could think of," says Peter. "They went right through Goldman Gates and even the individual more granular scorings were very promising indeed—we were particularly proud of that fact, that the companies would really move the needle. Until we learned, slowly but surely, that the size of the companies were perhaps a little too much. The needle could move too much. What you do in a situation like this is you revisit your individual scorings and add a new soft cap on size of company. If the shareholders think it's too big a mouthful you have to listen and learn. This process is not only beneficial for the M&A Deal Committee or teams doing the deals—it is also a way to align both shareholders and the management of the firm."

3 Communicative Leadership (Step 2)

What drives M&A success is not necessarily the opposite of what drives M&A failure, with the exception of one thing. People.

The final writing of this book took place in Belfast, Northern Ireland. We were trying to think of a few examples where people were responsible for the failure of a deal. We remembered the big bank merger where people aligned themselves with different colors, but we needed a stronger example. Then we looked out the window.

Obviously, the creation of Northern Ireland and its absorption into the UK cannot be compared with an M&A deal, but there is an interesting analogy there—that people can (and will) fight long and hard against something they don't like. Don't ever allow that to happen when you work with people and M&A transactions, both parties will fail and there will be no winners.

Of course, we are not saying that our M&A Formula can bring about world peace, but there are countless real-world examples of people being 'acquired' against their will, and fighting back against the 'acquirers.' If history has taught us one thing, it is that we need to respect the people around us if we want to avoid conflict. Make sure that the people who need to follow you post-deal *feel that they belong and are highly respected* throughout the process. That's the one thing you never want to fail on.

Lesson #1 in M&A Leadership: Drive Hard with Soft Management Tools

The whole idea behind business model-driven M&A is to secure full transparency for everyone involved and let everyone speak. If you ask a soldier, they will always be crystal clear about the purpose of their mission. We have met several external advisors, corporate entities, and a bunch of others involved in M&A who are just performing their roles in isolation without knowing the answer to that essential question: "Why are we doing M&A?" They will fail at least 50% of the time, as the organization is simply not supporting the deal.

When negotiations are going on, it is absolutely crucial to be upfront with the 'should be' business model, and to illustrate this thoroughly—and repeatedly—to the whole organization. Or rather, both organizations.

If the parties do not agree on the reason behind the deal and the eventual outcome, then it is often better not to do the deal. Unless, of course, you don't need their agreement—DSV, for instance, does M&A to acquire company activities and blue-collar workers. They are less dependent on management retention, as they can always provide their own.

The Goldman Gates process is a great way to get the whole team on board with the whole M&A vision. The 'go'/'no go' system tells everyone at a glance why a particular deal is (or is not) happening. Maybe you can communicate the Gates in an all-staff email, in a monthly meeting, or through your M&A Deal Committee (more on this later), but it is vital that it is communicated.

Many C-suite executives claim that they don't have time for communication, and prefer to focus on 'department matters' like tax, funding, and business development. Strangely enough though, they always have time to clean up after a failed M&A deal.

What we have learned from the M&A Elite is that they always punch above their organizational weight. Our favorite example is ASSA ABLOY, which drives so hard on M&A initiatives that its country heads and chiefs of products are expected to spend an allocated amount of money in order to deliver 5% M&A growth each year alongside organic growth 5%. You can also reach this level, and so can your company.

But you have to create a workplace where people feel valued and want to contribute, as opposed to being told what to do.

Lesson #2 in M&A Leadership: Create a Strong Foundation for Your People

Corporate Foundation

The seven companies in our M&A Elite represent seven potential 'role models' for any successful business. But they can also inspire you to take a long, hard look at the corporate foundation for your whole organization when doing M&A.

As we previously stated, there must always be a clear answer to the question: "why do we do M&A?"

All people like to feel belonging—so how do you create this sense of belonging among your people and the people you 'acquire'? What is their foundation?

Each member of our M&A Elite takes a different approach to this challenge.

FrieslandCampina created their own foundation post-merger by simply announcing a completely new CEO who then created a new logo with his newly appointed top-60 team. This may seem like a simple solution, but at the time it was viewed as nothing short of a Herculean task.

LVMH handles their corporate foundation in an indirect way as they insist on keeping the heritage of the brand. That way they not only avoid risking the existing foundation of their newly-acquired companies, but they can actually strengthen them. In particular, in identifying the top locations in the world's biggest cities, LVMH can position their new brands in a way that will gain them maximum exposure. As we previously stated, many Marc Jacobs, Thomas Pink, Kenzo, TagHeuer clients do not know that LVMH owns their beloved brand. In fact, many employees don't realize this either. That is exactly what LVMH wants to achieve, as they can keep each brand's foundation post-merger.

Another example of corporate foundation can be found at RB. The single most important driver for their M&A success is likely to be the very strong alignment between employees and shareholders. The idea is that they own RB and they want the organization to win and be successful with their takeovers of other firms. I'm happy to stand corrected, but my research shows that RB has delivered what looks to be the highest cost/revenue synergies, not only in their own industry, but across most industries.

The corporate foundation for DSV is cost, cost, cost. Having said this, they do deliver on many other building blocks and have received numerous prizes for best transportation, best supplier, etc. Still, in corporate M&A their corporate foundation is to reap cost synergies—preferably within 12–18 months. A corporate foundation is not always nice or comfortable for all involved. DSV had a period of almost seven years where they didn't find appropriate takeover targets as their business model driver is to reduce cost by closing down IT systems, laying off level-1 senior management, and streamlining transportation.

The corporate foundation in M&A is often built on the acquirer's successful business model.

What will your corporate foundation be in M&A? Do you prefer an indirect LVMH model, or do you firmly introduce your own business model firmly like DSV?

Whatever you decide, it is vital that you can define your corporate M&A foundation. Make sure everyone in the organization can understand it. Then act with great decisiveness, as your whole organization can now work towards the same goals.

Lesson #3 in M&A Leadership: Silent People Are NOT Team Players

Any leader should be able to communicate your M&A business model drivers, but that is only part of the M&A Formula. You have to ask yourself if the information—the M&A driver (or drivers)—is understood by the employees.

In order to create a 'sense of belonging' among your workforce, you have to talk to them, but you also have to listen. We can't overstate the importance of two-way communication, and creating an open workplace where it is easy to cascade information and get your message across. This is crucial in corporate M&A. You may think right now that this is an obvious point to make. Yet, in many M&A deals people forget that the speed of change is so much higher in 'buy' than in 'build.' When you build, there is more time to secure a feeling of belonging and involve human resources. The same cannot be said in an M&A deal. By the time you learn the real story about the worries of your colleagues or employees, it is usually too late to react. This is particularly true for the newly acquired people with whom you have no previous history.

There are a lot of soft factors on both sides in an M&A deal. Uncertainty prevails and people sometimes have a high level of ignorance. The relevant information is often filtered by C-suite and the M&A department itself. Due to time pressure in an M&A deal, individual employees are not heard. If their ideas are not heard, you may have a problem as a buyer.

One way to further enable the workforce is to use collaborative technology platforms (CTPs). In fact, digital development teams have been using CTPs—such as Slack and Facebook—for some time.

CTPs move beyond traditional stakeholder management in an M&A process and add cultural and social elements, which are important in assessing the overarching opinions of the workforce. It allows the leadership to ask simple questions like: "Are my staff happy?"

Organizational structure evolves into quite complex matrices and hierarchies. CTPs can provide capabilities to move a step beyond the conventional chains of command, breaking down communication paths into a network, rather than following a hierarchy.

CTPs encourage openness and have the capacity to bring forth information, previously buried in correspondence such as emails (bilateral communication), which can innocently be filtered away to never see the light of day.

However, there is one caveat. All of this comes with a willingness on behalf of the company leadership to be open. Silence is not a course of action. Silent people are not team players. Voices must be heard, and hidden agendas must be seen.

Through CTPs we gain the ability for greater workforce transparency and we give the workforce and company expertise a voice.

The voice can be used to assess whether a deal has potential challenges. Not because there is a lack of product market fit, or because the return on investment is low, but because there are cultural or perhaps technical challenges unknown due to poor communication channels.

Giving the workforce a voice also has positive inclusive effects. Companies are social organizations, especially in networked and matrix-based structures. Openness and transparency may also assist in mitigating any business model ambiguity.

There is evidence to support the fact that engaging a broader array of stakeholders in a company can generate a better foundation for trust. Interconnected digital platforms now allow everyone to engage seamlessly, in real time and across organizational and country borders, with more information than ever available before. This means that you are able to make informed decisions and build trust at scale. It also has a compounding effect, accelerating the rate at which M&A can happen for you and your company (as well as your competitor, so don't ignore this development). You'll learn more about mastering the use of these platforms to gain trust and increase your success rate.

Invest in the success of those around you, using digital as an enabler.

The M&A Deal Committee

Everyone knows that an octopus has eight arms,[1] but what you might not realize is that this creature is highly intelligent and normally gets what it wants. It is a fantastic hunter and ferocious killer in the animal kingdom. In fact, the brain of the octopus is far more advanced than the normal human brain system. Okay, we Homo sapiens are much smarter than octopi and the human brain is thought to have about 90–100 billion nerve cells (neurons), where the octopus has around 500 million. However, these neurons are split across nine different areas—the central brain, and each one of the eight arms.

Recent research has found that any one of the arms can still work independently, even when it is not attached to the body of the octopus. What's more, the central brain only uses one-third of the total neurons, with the majority of nerve cells being delegated to the arms.

This is perhaps the best way to describe an M&A Deal Committee. First of all, only about one-third of the corporate neurons for M&A are located in the M&A department itself, the rest are outsourced to the various other arms of the company.

Putting Together an M&A Deal Committee

The 'arms' of an efficient M&A-savvy corporation might include:

- Business Unit (the acquirer and the owner of the business model!)
- Corporate Strategy
- Business Development
- Treasury
- Tax
- Finance department
- Legal department
- Procurement
- and, of course, the M&A department.

Each of these individual arms is crammed full of highly specialized knowledge, and each one is equally essential to the success of the deal.

Every M&A professional understands the value of this extended team. After all, how could an M&A expert be expected to know about the finer details of commercial taxation, or the legalities of acquisition in an unfamiliar country? The most successful Deal Committees will surround themselves with high-powered arms and let each 'brain' do what it does best.

Communicative Leadership (Step 2)

The M&A Deal Committee is made up of leaders—department leaders, thought leaders, and M&A leaders—who all act together in a collaborative network.

So, before we familiarize ourselves with the Deal Committee, and the role of the various arms, it is worth taking a moment to remember the value of great leadership.

The purpose of these business units is to activate the *whole* organization and increase flexibility, so that the company can act quickly and decisively on any M&A opportunity. Don't forget that a quick rejection of a potential project (which will be added to your M&A anti-portfolio) is a clear sign of corporate M&A competency. Great M&A deal-making is not only about making deals, but also about identifying the deals you don't want to do because there is no logical business model fit.

Remember, half of M&A success lies in the deals you never do. With an apparent global failure rate of at least 50% in corporate M&A (although we have proved that this myth is often wrong on a company basis), most companies need to learn to say no a lot more often, and preferably before they have spent a lot of resources on the wrong projects.

Creating Strong Alignment of C-suite, Deal Committee, and Organization

Gather together a group of C-suite employees at a corporation, and they are unlikely to be 100% aligned when it comes to doing M&A and what the company should be looking for. If the most senior managers can't agree on why or how they are pursuing a particular deal, how can they expect their 'arms' to work together for a common goal?

By using business model-driven M&A you create alignment, and further establishing an M&A Deal Committee creates accountability and action when needed. These C-suite managers can delegate M&A responsibilities to specialized business units and divisional structures. However, you cannot delegate without being crystal clear on what you want. C-suites who keep M&A projects entirely for themselves, up in their ivory tower, often do not know exactly what they want. They think they are aiming for the bullseye, but they are throwing the dart first and drawing the circle afterwards. Without providing empirical studies, it is likely that many deal failures will be rooted in bad M&A leadership—just like the case of countries 'acquiring' other countries.

Communication is essential at every level of an organization. There's a reason why 'leadership' is number two in our M&A Formula. Without good leadership, the formula doesn't work.

Corporate M&A is not only about valuation and negotiation tactics. When it is successful, it is because you were able to activate the entire organization by making them act on their own, like the arms of the octopus.

CEOs will generally be quite resistant to this strategy, as they won't want to divert their resources and distract heads of departments by bringing them into endless meetings. However, in practice, you will win back these resources twofold, as you turn those departments into brand defenders of M&A.

No company is too small to justify an M&A Deal Committee. The committee can be made up of 15 seasoned executives, or two partners from different ends of the business. It really doesn't matter, just as long as you are all bringing your individual expertise to the table in service of the same goal.

As with most things in life, M&A is about people—the people in the Deal Committee, the people in the acquiring company, and the people in the target company. So, what happens when you want to buy a company because you badly need some of their business model 'building blocks' for your own business model (e.g. IT tools as a new channel), but you realize that the complementarity between the two companies is frighteningly low?

These are the 'handle with care' deals. You should not necessarily avoid these deals, but you have to be aware of the impact of incompatibility. For instance, you may want to treat the newly acquired company more as a partner than an acquired asset, "Don't integrate your acquisitions, partner with them" being the name of an interesting article from 2009.[2] However, it is still valid today, not least because of the issues which occur when there is low complementarity in the Goldman Gates Scoring. The article seems to be inspired by Leo Tolstoy, regarded as one of the world's greatest writers, who proclaimed that happiness in a marriage is not really about how compatible people are, but how they deal with incompatibility. What the authors refer to in the article is allowing acquired companies to continue operating independently, almost as if there had been no change of ownership.

The authors refer to this type of integration as "light-handed M&A style." We just call it integrating fewer building blocks into the M&A deal.

LVMH are the experts in this field. They keep every acquisition structurally separate for some M&A drivers and maintain the newly acquired company's own brand and identity. They even allow each company to maintain its own value proposition with regard to client relations and customers. In the eyes of the employees and customers, nothing has changed, except for the fact that the new store might be located in a higher-profile area. Cross-border mergers and acquisitions are notoriously more difficult than domestic deals. The problems can get even bigger when there is a low business model complementarity, according to academic research. LVMH's approach has proven to work brilliantly. The building blocks that can become issues are simply not integrated into the business model, whereas Customer Perception remains the same, as well as Customer Retention and Value Proposition.

The most successful stories in corporate M&A come from organizations who actually believe they can do it. It is a self-fulfilling prophecy. When an organization has a fear of M&A, this will trickle down to all members of staff. But if you think you can do it, and you can see a direct route to success, then every other employee will be on board and support your M&A projects.

The Business Unit Approach

An M&A Deal Committee could also be used as a kick start for a company that wants a new beginning in M&A. In fact, some of the companies used as examples in this book have adopted a similar approach as a way of activating the whole organization.

One commonly used organizational structure in M&A is to run projects in various divisions—typically called business units. Such units can be product lines, geographical responsibilities, or sometimes even a combination of both.

What works best for your organization will depend on your internal competencies. What do you have in-house with regard to M&A expertise? What is your M&A foundation? Does your M&A team and all the 'arms' supporting it know exactly what game you play? What business model fit are you after? When do you need to deliver those synergies and how likely are you as an organization to reach those goals? Do you have people to whom you can actually delegate all these tasks without the need to be there in person as CEO?

In a sense, your M&A Deal Committee is just another team, or 'unit,' within your organization. These are the people with the skills needed for the specific task of finding new opportunities to be added to your business model, and as such they represent the whole organization. Most companies will be too bureaucratic in their existing organizational structure, and important information is likely to get lost. Most successful M&A firms—or at least the Global M&A Elite—use dedicated M&A teams to perform business model-driven

M&A. The team will score each M&A project, run the project, constantly observe and share any information closely, and report to the C-suite.

The more deals you decide not to do, the more anti-fragile you will become. In doing so, your M&A Formula will become leaner and more anti-fragile.

In organizations with a divisional M&A structure, there is still a central unit which will delegate authority. In some cases it will even allocate M&A 'war chest' money to these divisional structures, with the message "go get new companies into our business model."

ASSA ABLOY is a famous example of this, forcing each department to invest in new companies every single year. To some business-owners, this will seem pretty scary, but that is only because there is an inherent lack of trust, or a lack of deal-making experience.

The 'Business Unit' approach cannot be applied to companies which do not have the necessary practical experience. It is hard to achieve the required level of practical experience by doing only two or three deals a year. These companies may be better off with a formal M&A Deal Committee supported by a strict M&A Launchpad, complete with formal decision gates and support from HQ with regard to the M&A dos and don'ts. After all, if you are learning how to become a trapeze artist, you don't start by jumping into thin air without a safety net.

When you get more practical experience, you may be able to act on your own. For instance, hunting specific targets or operating within a well-known area. Business unit leaders have to deliver both the 'build' and 'buy' options.

Applying M&A Goldman Gates Scoring

Figure 3.1 shows an example of Goldman Gates Scoring by an M&A Deal Committee. It is based on a real-life business model fitting,

Which are the two business units which most clearly manifest the CORE COMPETENCE of your firm?

(CORE COMPETENCE:
1. Provides potential access to many markets.
2. Makes a significant contribution to the perceived customer benefits of the end product.
3. Is difficult to imitate.)

Acquiring Business Unit:

Acquired Business Model:

What are the main products?

Acquiring Business Unit:

Acquired Business Model:

Please compare existing business model (acquiring firm) and new business model (acquired firm). To what extent are they similar as regards to the following

COMPLEMENTARITY SCORE 4.375	Very Different		Medium		Very Similar	
Product-Market Building Blocks	1	2	3	4	5	Comment
1 Product Use					5	
2 End Customer Types					5	
3 Competitors				4		
4 Product Design				4		
5 Product Technology					5	Firm higher RD/Sales + technology leader
6 Pricing				4		
	0	0	0	12	15	27
						4.5

Communicative Leadership (Step 2)

Resource Building Blocks

7	General Management Skills				4		
8	Technical Skills					5	No economies of scale prior to M&A
9	Administrative Skills			3			
10	Brand Identity					5	Domestic target
11	Brand Recognition				4		Domestic target followed by dual branding
12	Market Knowledge					5	
		0	0	3	8	15	26
							4.3

Value Chain Building Blocks

13	Sales Channel Type					5	Firm has direct and in-direct - all channels
14	After-Sales Services				4		Firm more up to date 24/7 approachable
15	Supply Channel Types				4		
16	Suppliers				4		Suppliers are replaced with own products!
		0	0	0	12	5	17
							4.3

Figure 3.1 Questionnaire to corporate executives in the Global M&A Elite

but has been moderated in order to avoid compromising any firm in this book.

Assume a company has identified three building blocks in the business model that need to change. Management can either change them over time or use M&A to enact these changes more quickly.

Together with the Executive Board, the M&A Deal Committee sets five main priorities to look for in M&A origination. Companies like DSV, Danaher, and the global brewer are unusually well articulated in what they are looking for, and rank as among the best in the world when it comes to business model-driven M&A.

The first thing to consider when you approach Goldman Gates Scoring is the generic business model drivers.

Driver #1—Geographical Fit
Does this target deliver the geographical expansion you are looking for, and to what extent?

Driver #2—Value Proposition
Does this acquisition support your existing portfolio with power brands?

Driver #3—M&A Value Drivers
Do you have the key resources and key activities needed to support future growth and scalability?

Driver #4—Market Position
What is this company's market share, gross margins, and current cost structure? And can you change this for the better?

Driver #5—Overall Business Model Fit
How do the two business models fit, and is the seller willing to take this journey together with you?

It is crucial for a company like DSV, Danaher, or any other member of the M&A Elite that all business owners have a 100% buy-in to the newly merged business model. In some cases, top-notch M&A Elite companies will simply turn down an acquisition opportunity because they have doubts about the management's intentions. For instance, DSV is known for having high company morale, where every member of the team is aligned with the company vision. But pre-acquisition, many of the top-level executives are often fired and anyone who is unhappy with the merger will leave. Sometimes you have to be brutal to reach your M&A goals.

There must be a business model fit that both parties agree on. The top seven M&A Elite companies have seen success because of their commitment to this approach. Of course the world of M&A has hostile takeovers, but it is normally easier to cooperate and with that comes a much higher chance of success.

The next thing to consider is the target assessment. This matters for a number of reasons.

Financials. The size of the transaction is important to many firms, as an M&A deal should always 'move the needle.' In some cases, a costly and time-consuming acquisition will only buy 5–10 business days in revenue. While it is important for all companies to follow strict M&A procedures no matter what the size of your target, it is worth remembering that there may sometimes be a disproportionate amount of time spent on smaller targets compared with bigger targets or even game-changing deals.

When it comes to financials, it is useful to have a critical review of the expected cost and revenue synergies. One of the benefits of following these M&A preparations is that they are better at consolidating two business models into a new 'should be' firm. Investment banks do not really care if you are on a realistic mission, because by the time any cracks have started to appear, they will already have moved on to the next deal. A typical range for cost synergies to sales is perhaps 2–10%, although some members of our M&A Elite (RB, for instance) have reached levels much higher than 10%.

In the final scoring, you will take all of the financial parameters and come up with a 1–5 figure on the expected financial attractiveness of the target.

There are exceptions to this process. ASSA ABLOY is one of the few companies which has a commitment to buying even the smallest available company to make potential sellers feel as though they can approach the firm when they want to sell their business. For some companies this would never work, because they do not have M&A processes set up for smaller acquisitions. This should then be reflected in their Goldman Gates Scoring. A deep dive into ASSA ABLOY's acquisition history for the last 10 years found that 75 companies (out of 150 companies acquired in total) accounted for more than 90% of all the company value acquired. Many companies would not waste time buying the 75 companies which yielded below 10% of the total acquired growth, but it all depends on what game you are playing and what M&A reputation you want to build.

Resources. Even ASSA ABLOY will sometimes pay more attention to big-game hunting than very small transactions. Particularly if most of the team is snowed under thanks to due diligence and other bureaucratic processes.

Every company must cleverly choose its M&A resources, and maybe even increase the number of people hired from buy-side advisors. The important thing here is that the M&A Deal Committee must be aware of their resources before any deal-making even begins to take place, so that they can avoid chasing M&A projects they may not be able to follow through on.

Feasibility. Do you buy a team or a one-man army? Based on your discussions with the target firm, what have you experienced? Have you identified 'key man incident' issues?

With every deal, there is an opportunity to take stock of the total cost, both in terms of the financials and in terms of your commitment. When do you conclude that enough is enough and start looking for something more do-able? It's important to set these parameters right from the very beginning. Some M&A deals take a very long time to complete, and you don't want to be waiting forever for a deal that isn't a good fit.

Funding. In practice, it is useful to separate the total funding involved in running the business and the M&A projects. Why? Banks are inherently known for lending out umbrellas when the weather is good.

Once the storm clouds appear, they want their umbrellas back for their own protection.

There were a lot of casualties during the global financial crisis of 2007/2008, as the banks either didn't extend corporate loans or simply discontinued them. It is therefore crucial that any acquisition can be carried out without taking funding away from running the business, and thereby putting the company at risk, should there be another clampdown from the banks.

During the financial crisis, the top business model-driven firms were able to secure funding by agreeing their entire M&A strategy with the banks and capital markets.

For instance, FrieslandCampina uses M&A bridge financing. But while most companies will agree bridge financing over a 6–12-month period, when I was at FrieslandCampina I worked with the Treasury department to secure a 5-year M&A bridging loan from the banks. We did this because just a few years earlier, we all saw the effects of the financial crisis—short-term business loans were among the first to be withdrawn or discontinued when the banks were under pressure. If you rely on capital market or bank funding to do M&A, you are at the mercy of the economy for the duration of the loan. A lot of volatility can take place in 6–12 months, whereas a 5-year period offers a lot more flexibility. In our experience, a combination of longer-term funding from both banks and investors is absolutely crucial for your M&A success. When it comes to choosing banks for both funding and financial advisory services, it is advisable to let your Deal Committee do this, rather than leaving it to one or two departments. The Treasury department will normally lead this selection process, but it is often assisted by other members of the Deal Committee to secure an anonymous vote for the bank you choose as M&A advisors. It is important for your company that no one, including the CEO, has a personal bias with any financial supplier whatsoever. In my time on the M&A buy-side I would never accept a dinner reservation or any other form of 'kindness' from an M&A sell-side above the value of €100. We would also never take part in any meeting arranged by sell-side suppliers, unless it was clearly a work-related topic. Most corporates demand that the procurement of all products and services be carried out, or at least assisted by, their Procurement department. Why would that not include the procurement of M&A financial services?

Intent-Based Leadership

> *Bad leaders give bad orders; good leaders give good orders. Great leaders give no orders* (read the whole story in *Turn The Ship Around!*).
>
> DAVID MARQUET, FORMER NUCLEAR SUBMARINE COMMANDER

David Marquet took over USS Santa Fe when it was the worst-performing submarine in the US fleet. However, he turned it around by moving from a leader–follower style of management to a completely new method which he called 'intent-based leadership' (see more on www.davidmarquet.com).

This theory was born when, a few weeks into his new job, he gave his crew an impossible order during a drill. Asked to simulate a fault with the sub's nuclear reactor, Marquet ordered the helmsman to turn a small motor "ahead two thirds." The crew obeyed, and nothing happened. When he asked why, he was told that there was no such thing as "ahead two thirds" with this particular motor, yet they were prepared to follow his orders regardless of whether or not they were technically possible. In a real-life situation, this could have been catastrophic.

When it came to the specifics of the ship, the crew knew more than he did. Of course they did; they'd been working on the same ship for years, while he had only been there a few weeks. He decided to make a big change—turning the crew from followers into leaders.

In a very short space of time, the USS Santa Fe went from worst to first in the Navy. The submarine went on to become a learning organization, and continues to win awards and promote more officers than any other submarine, including ten subsequent submarine captains.

Marquet's leadership style was the inspiration behind the M&A Deal Committee. Many companies fail to take responsibility for M&A, or concentrate all of that responsibility at the top. Meanwhile, traditional managerial approaches simply do not work in many companies.

We asked David Marquet to visualize his 'intent-based leadership' in relation to corporate M&A. He said:

When the business owner or CEO is making all decisions, the company is a much less attractive target because you're buying a person not an organization.

In cases where this is true but not visible to the acquiring firm, it's often a disaster because that person is living on their own island and the company is most often biased towards passivity and inaction.

Firstly, this shows the importance of scoring M&A feasibility in Goldman Gates (particularly with regard to the key man incident risk). Secondly, this is what happens in an organization that has not implemented M&A broadly in the organization but is cascading deals from the CEO ivory tower.

Notes

1. Some claim only six arms and two legs. Other octopi have ten arms, of which two are tentacles. But that's not the point we are trying to make here…
2. Kale, P., Singh, H., & Raman, A. (2009) Don't integrate your acquisitions, partner with them. *Harvard Business Review*, December: 109–115.

4 Take Ownership (Step 3)

Once you have identified and pursued your business model drivers, communicated your M&A vision to the entire organization, and ranked each project in the Goldman Gates, it is time to get to work. And you have to act quickly, before the opportunity to gain has been lost.

The M&A Launchpad is there to help you either reject or complete your next deal as quickly as possible.

Remember—be brutal. You are always looking for a reason to eject each candidate at any given stage, so that you can catch any potential pitfalls before they cause you serious harm. Don't allow emotional reasons to cloud your decision. M&A is not an emotional task, it is purely business model-driven. Either a deal works, or it doesn't.

Next to 'Really Bad' deals (done deals that failed miserably), there is nothing worse for shareholders than 'Bad' failures, which we define as deals which had the right business model fit in Goldman Gates Scoring, but were lost due to poor M&A practice in the company.

If you are a member of the Global M&A Elite or one of the high performers among SMEs, losing a deal due to poor M&A practice is a very painful process. To take the example of my favorite tennis player again, it would be like Serena Williams arriving on court without knowing whether she is playing a singles or a doubles game.

In the Introduction, we compared the world's top fund managers (indirect investments) with corporate M&A (direct investments). There are many similarities between these two types of investors—the overarching mission in both cases is to invest cleverly and get a high total return.

In the world of top fund managers there is not much time to make decisions.

Weeks or months of waiting time in M&A is like minutes or seconds in the world of capital markets. When a top investor sees an opportunity, there is no time to go through long discussions internally before making a move. Compliance functions are there to *support* fast decision-making, not to prolong the process and jeopardize deals. Before any deal goes ahead, the fund manager will already know how much money can be invested in each security type depending on risk, currency, maturity, and any other factors.

You may never feel this kind of rush in corporate M&A, but it is increasingly happening and for the very best companies there is an intense rivalry to do the best deal first. They know that they can get to the table by bidding high, and they can reduce risk for the seller by offering them attractive terms or even paying M&A insurance. You need to make your company an attractive buyer; at every step of the way you have to constantly make the seller feel comfortable with your process. For instance, you can reassure the seller by showing them security of funds at an early stage.

Never ever lose a wanted deal because of poor M&A practice. Your M&A Launchpad is there to give you immediate support, so you can act quickly when a deal is 'go.'

Take Ownership (Step 3)

The M&A Playbook

Wherever possible, corporate processes need to be in place so they can be drawn upon as soon as an opportunity is presented. This sounds boring, but it's more important than you think. There is no point in discussing M&A processes while you're in the middle of the deal—you don't see Serena Williams chatting to her trainer as she lobs the ball back and forth. Too many deals don't come through because the buy-side was unprepared in some way, and the opportunity to gain was lost. That is unacceptable. That is a 'Bad' failure (see Figure 4.1).

If you have bad practices internally and you cannot make quick decisions, it doesn't matter how well your business model drivers line up—you will lose.

The M&A love letter is often the first step in an M&A deal (or letter of intent [LOI], memorandum of understanding, or whatever other term your company uses to describe it). This is a very important step for an individual target.

When it comes to decision time it must, at all times, be the legal counsel who formally performs all the cross-checking and reporting on

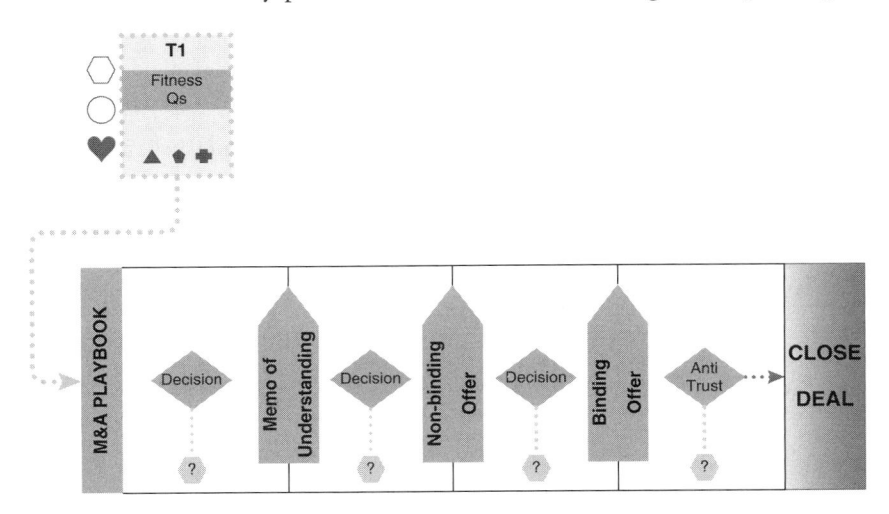

Figure 4.1 M&A Playbook (see also www.fixcorp.co)

the buy-side. It may be useful to have security of funds before actually considering a non-binding offer to a seller. There is no point talking to someone who cannot raise the money.

Most companies have internal rules as to who can actually sign the LOI, non-binding offer, and binding offer. Somewhat surprisingly, these rules often do not support what I consider to be a highly effective M&A Launchpad. Most companies are afraid of M&A, and so they find comfort in tightening up the rules and procedures in their M&A Playbook. All that only slows down the M&A process, and will not actually help you avoid the 'Really Bad' deals—deals that destroy careers and lose shareholder value. They are the totally unacceptable failures in M&A. At best, a clever seller may take advantage of a poor M&A execution team, believing them to represent a poor company in terms of M&A capabilities. In other cases, if the acquisition launchpad and internal rules are not pre-made agreements both internally and externally, your organization may lose deals that were highly desirable because the buy-side was simply unprepared. That's what's called a 'Bad' failure.

Companies who are acquiring hundreds of firms cannot rely on tight legal or compliance systems in M&A. Serial acquirers must rely on their own organization to do relatively smaller deals. For instance, mandating country managers to be able to sign deals up to a certain threshold, just as long as they follow a rules-based system (business model-driven M&A).

If you want to transform yourself to become successful in M&A, you have to combine business model-driven M&A with a strong M&A Launchpad.

Governance in Corporate Processes

Most companies have three or four decision gates when they handle single M&A deals. Some companies call these decision gates 'M&A traffic lights,' or simply refer to their M&A Playbook.

The first decision to be made is when you sign a confidentiality agreement to give you further insights into a new target or get approval to approach a new target unsolicited. If you act on a target that has been set up for sale, make sure you react immediately and get the information ASAP. The use of an internal M&A Deal Committee can speed up this process, and if they use Goldman Scoring, they could be in a position to make a 'go' decision within 48 hours, which includes hiring an external advisor! The Corporate Framework Agreement (CFA) (see more on www.fixcorp.co) speeds up your process, as it means that the confidentiality agreement, terms and conditions, and even the fees have all been arranged with your financial advisors from the outset. You just have to drop the M&A assignment in with your M&A panel of banks as per the Request for Proposal (RFP), and your own Deal Committee will appoint an advisor within 48 hours. This is a process that sometimes takes one or two weeks or more, and is akin to a Formula One driver wasting time in the pit.

The next level is often the LOI stage (which may or may not include a possible exclusivity for the buyer), followed by the non-binding offer and binding offer, and finally closing of the deal. The Standard Operational Procedure (SOP) must follow this, to support fast decision-making. The more your team can deal with upfront, the better.

Agile funding works the same way—successful M&A firms keep their gunpowder dry and know exactly how much firepower they have available for M&A deals. In some cases, they will separate the funding needed to run the business and the funding to be used in M&A activities. The banks supporting M&A deals with M&A bridge financing (preferably 3 to 5 years long) may sometimes have limits as to how much can be bought by the company in certain regions or countries. Any potential limitations must be cleared immediately by the Treasury department.

The Tax department may have a preference for what kind of deal structure is optimal for your company, and sometimes they may even suggest using royalties and other tax-optimization tools to make the deal more attractive.

Other departments which may be included at the early stage of a deal include communication and HR functions. These are more important than you might think, and include the people who will help your team understand each element of every deal. If you want a high retention of people in the company you buy, you need HR to put together a retention package to make sure nobody wants to leave you during the takeover of the firm.

Finally, the Legal department will be the boss when it comes to handling the M&A governance and as we will see later, companies may benefit from building a so-called M&A buy-side catalog with professional, highly reputable, M&A legal advisors to build a strong and solid M&A Launchpad.

Run these steps every time you are considering a new acquisition, and don't proceed unless you have met every requirement.

The M&A Formula does not change, and it doesn't have to. Follow these steps and they will lead you towards rules-based investing or what we call the 'Goldman Gates,' where you will make immediate 'go'/'no go' decisions.

CFA: Company-Specific M&A Engagement Letters

"I remember a discussion that took place at an M&A council meeting, about the long and winding road to hiring an investment bank for an M&A project," says Peter. "What is a reasonable success fee? What do

Take Ownership (Step 3)

you pay as monthly retainer? Should you even pay a monthly retainer at all? But one issue that stood out from all others with absolutely no variables or flexibility was 'the holy M&A engagement letter.'"

"It was clear from the conversation that everyone agreed that it was impossible to change the way M&A engagement letters were formulated—it would also be insane to change the way things had always been done when it came to hiring banks for M&A jobs. That was the moment that I knew I had a problem with a binary outcome: either I would make the case that we needed a complete makeover of the M&A engagement letter; or I would just accept things as they are. The second option just felt like giving up. It is apparently my curse in life to always strive to optimize things of whatever kind. I really cannot help it."

"However, these opportunities should always be considered with a 'Pareto optimality' approach. There is no point in making yourself much better off if it leaves others much worse off, as Pareto optimality is about making everyone better off without making anyone worse off."

"I felt certain that my approach to changing M&A engagement letters would be a win–win for everybody," added Peter. "After that council meeting, I felt brave enough to challenge the long-standing tradition of just blindly signing the M&A engagement letter from the almighty M&A advisor."

"I told my colleagues that I was going to change this practice completely. Within a year, I would develop a brand-new procedure for hiring banks and at the same time stop using M&A engagement letters and make my own company-specific M&A engagement letters instead."

Peter claimed that this solution would make the M&A process so lean that it would compare the in-sourcing of investment banks with the Procurement department's procedure for buying A4 print paper.

Not only did all of this happen, but a firm in the same league as the M&A Elite was able to reduce its M&A costs by more than $1 million per year—money that they would have spent on financial advisors. Add to that a much more effective sourcing system and dramatically reduced execution risk in M&A.

"Looking back, I can't believe how much time I once wasted in my former role as head of M&A," Peter adds. "Negotiating terms and conditions with your own advisors, bartering over the monthly retainer and other non-value adding stuff; all of that time could have been better spent finding and making deals!"

"I knew the message was sinking in when a procurement colleague began referring to one of the world's top-10 investment banks as a 'supplier.' In his mind, 'buying' advisory services from them was akin to buying A4 paper from Canon."

You have to completely remove the mystique around sophisticated financial products, and define the world in terms of simple 'goods and services' instead. You cannot miss any opportunity to look for a newer, better deal from external sources. By treating banks as suppliers, you will immediately be able to consider the most appropriate service, in terms of quality, quantity, time, and location. This will allow you to promote open and fair competition, and ultimately get a better deal for everyone involved.

M&As are all about execution—doing the deal, making money, and moving on to the next challenge. Anything else is a waste of your time.

The only way to maximize your million-dollar savings, streamline the process, and execute the deal is to make your own M&A CFA that is commensurate with your strategy, indemnification limits, liability needs, and termination rights. You dictate the Ts&Cs and the pricing, not the supplier.

One person who has had a huge impact in the development of the M&A CFA is John Schultz. A former M&A lawyer at Allen & Overy, Schultz spent a number of years working with investment banks and their clients, helping them to put together M&A engagement letters. When he moved to FrieslandCampina, he met Peter Secher and they realized that they could combine their skills and save money for the firm by developing a company-specific M&A engagement letter with best practice in all areas.

The CFA concept started with financial advisors. In the last few years we have broadened the corporate framework to include legal advisors and transaction service providers (TSPs) in any M&A deal.

What Is the CFA?

The CFA is a framework of agreements set up between you and your external M&A advisors—they can be banks, accountants, or legal advisors. The combined set of agreements are pre-qualified and adhere to terms which benefit any corporate who has M&A ambitions and wants to act fast, cost effectively, and with the most predictable outcome of success.

The three main stages to CFA are:

1. Establishing a 'stretch goal.' Ideally, this will be done at the very earliest stages of any M&A Launchpad, but it can be reviewed at any time.
2. Establishing a cross-functional team. Preferably you will use your own M&A Deal Committee, which involves all disciplines of the business.
3. Empowering a CFA team to start implementing the templates in your M&A Launchpad and keep a unified communication line with C-suite executives.

Each company's CFA will be different, but the outcome will always be the same—faster, better, and more cost-effective M&A routines.

What Is the Engagement Letter (EL)?

"How many times have I heard companies telling me that they have never had any legal problems with engagement letters," says Peter. "And how many times did I tell them, that I have never had a car accident—in 30 years I have never once had to file an insurance claim for driving into another car. Clearly, I have been wasting my money on car insurance all these years, and wasting time buckling my seatbelt several times a day."

Except that we all know that we take these precautions to protect ourselves against the small chance that we may wind up in an accident which has the potential to change our lives for the worse.

When it comes to M&A legal challenges, the statistics are out there. If you haven't been hit yet, it's probably just a matter of time… You had better keep on buckling your seatbelt, just in case.

All companies can significantly limit the risk of an indemnity by following the best practice as laid out here. Companies that are happy to just leave things as they are could find themselves facing several costly cash-outs 'promptly upon request' (contingent liabilities). This does not happen very often but when it does, the instant repercussions are really, really bad.

You might think that your investment banks or financial advisors would handle any liability towards their corporate M&A clients, but the reality is that none of them are actually secured in any way. This is not best practice. In our view, external M&A advisors must take responsibility for their work and be liable for any mistakes (all losses, claims, damages, and liabilities). Without any kind of liability, an external advisor such as a bank, accounting firm, or boutique advisor may not be motivated to deliver the very best quality and commitment in high-pressure situations—for instance, where they have other big deals underway or where they have lost key personnel.

Here are a few things that you need to be aware of before signing any engagement letter.

Legal Considerations. Why is legal so important in M&A engagement letters? Easy—it reduces the M&A risk, saves money, and allows you to focus on the deal.

M&A litigation risk is higher than ever. In the USA, M&A filings have exploded since 2007, and it won't be long before the trend moves across the water to Europe.

In 2013, plaintiff attorneys filed lawsuits against 94% of all M&A deals announced that year and valued at over $100 million. In many cases, multiple filings were made against particularly significant deals. These include:

- Dell Inc. buyout (26 filings)
- Tellabs Inc. buyout (16 filings)
- Avago Technologies Ltd/LSI Corporation (16 filings)
- LinnCo LLC/Berry Petroleum Company (14 filings)
- Management buyout of Dole Food Company Inc. (13 filings)
- Toray Industries Inc./Zoltek Company Inc. (13 filings)
- KKR & Co L.P./KKR Financial Holdings LLC (13 filings)
- J. Heinz Company buyout (12 filings)
- Goldman Sachs & Co/Ebix Inc. (12 filings)

In the vast majority of these cases, shareholders were behind the filings. In fact, in 2013, 94% of M&A deals were challenged by shareholders, and 88% of all claims were eventually settled.

The odds of an M&A deal avoiding litigation are extremely low, so it's time corporate lawyers took action and made a CFA part of the M&A Launchpad.

Pricing and Fees. Most companies will blindly follow the lead of their financial advisors, TSPs, and legal advisors when it comes to pricing, and this is exactly what they want.

The best M&A buy-side companies will operate with no retainer and a 0.3% transaction fee (aka 'success fee'). However, we do not always recommend that you squeeze them too much on the transaction fees, as you must never, ever compromise on the quality of external M&A advisors.

It is also a good idea to look out for tail fees. The best in class will typically eliminate the tail fee entirely, but best practice is usually 6–12 months.

Termination Rights. Many contracts will include a caveat that the buyer cannot terminate immediately if there is a major change in personnel at the advisory firm. However, it is not unusual to see financial advisors being headhunted by competing firms. In fact, on Wall Street, in London, and in every major financial center, firms often switch both people and whole M&A teams. That is why we always recommend including a 'key man incident' provision for you as buy-side M&A. You have to be able to terminate immediately if your key advisor in the firm has left.

Furthermore, corporates should always include a provision which protects them from indemnity if an advisor terminates.

Other Terms and Conditions. These will be different for every company and every contract, but there are a few things to look out for.

First of all, be aware of where the deal is taking place. When it comes to governing law, New York is probably the most expensive place to litigate and potentially the most inconvenient. Best practice in global M&A would see all engagement letters filed under a single governing law, in an environment which is more business-friendly and less litigious.

Some ELs will also include reimbursement clauses which allow advisors to claim obscene amounts of money in 'individual expenses.' As far as we are concerned, best practice is to reimburse the advisor in line with the company's own travel policy (for instance, economy class for short flights and only business class for say 6 hours or longer). It is also worth adding that any costs made by third parties (e.g. legal costs) will not be reimbursed.

Codify Your Documents. Everyone in your M&A team should have access to all your resources and documents, and they should be filed in an easily searchable way. You want your people to work together, so encourage open communication and welcome any questions and advice.

M&A-specific software can make this process much easier and much more secure, so it is worth considering an investment in an online platform. And don't forget to include training for all new staff members—these resources are useless unless everyone can access them.

Digitizing M&A has become more and more popular in recent years, and it is a great way to kick start the implementation of the M&A Formula and accelerate your M&A processes. We will take a closer look at this in Chapter 5.

M&A Launchpad

Good M&A begins and ends with good governance. No deal should ever put your company and shareholders at risk, so your governance team should have the final say on the construction of the M&A Launchpad and how it is deployed (see Figure 4.2).

The Launchpad offers a clear set of instructions to make sure that the goals and purpose of any M&A deal are aligned with the shareholders' best interests. At this stage, it is critical to ensure that the reasons for attempting your next acquisition align with the goals of the company, and that no deal ever compromises shareholder value.

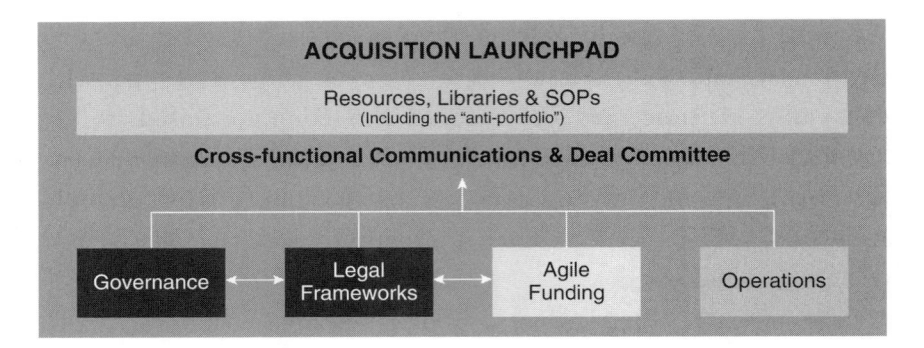

Figure 4.2 Acquisition Launchpad

Spend some time working through your M&A Launchpad so that you can create the strongest foundation for M&A agility and success.

M&A Legal

Corporate lawyers spend a lot of time working on contracts. These contracts might be about selling/supplying their company's goods, procuring goods and services for the company itself, looking at property interests, government regulations, patents—you name it, there is a legal discipline dedicated to it, and a million corporate lawyers who excel in the subject.

> *Top Legal Disciplines Ranked Annually by FT.com*
>
> - Business Development and Knowledge Sharing
> - Branding
> - Driving Value for Clients

- General Legal Expertise
- HR
- M&A
- Regulation
- Resourcing and Efficiency
- Restructuring and Reorganization
- Securitization
- CSR
- Speed and Process
- Strategic Collaboration
- Technology and Data Analytics

Corporate lawyers tend to be really good at what they do, but very few of them are experts in M&A. In a global firm, you will have dedicated people in your Legal department for M&A projects, but that's quite unusual for SMEs and even in some large corporates. Yet every once in a while, a company's legal counsel may be asked to assist in the acquisition of a company or a corporate merger. When this happens, it is time to seriously consider the role of the legal counsel or the in-house legal team.

The very best legal advisors understand how specialized law practice has become.

When met with an M&A legal challenge, every in-house legal team should consider two very human goals. Firstly, to know things. Secondly, to know who knows things.

The second one is by far the most important point when it comes to M&A.

M&A Legal: Don't Be Cheap! Any global corporate or SME owner should always consider in-sourcing their M&A lawyers without hesitation. If you think these M&A law experts are expensive, then try to do an M&A deal without a lawyer, or with your existing legal counsel who is simultaneously working on all the 14 subcategories listed above.

A key part of our M&A Formula is helping you to drastically reduce your costs for external advisors. We find it very strange that so many large corporates and SMEs spend the lion's share of their external deal expenses on investment banks and financial advisors when they should be spending so much more on their legal advice. As we have mentioned many times before, none of these external advisors will ever be able to help you reduce your M&A failure rate and (even more importantly), they will never ever make you excel and become a successful corporate acquirer.

You need business model-driven M&A. Still, M&A legal advice is highly valuable in corporate M&A and too many business-owners may not want to spend enough money on good M&A lawyers.

Case Insight: Mediclinic and Al Noor

In 2016, the *Financial Times* put together a list of the world's most innovative M&A lawyers, pointing to real-life cases as evidence.

Slaughter and May won the overall award for their work with a South African company called Mediclinic International. Mediclinic wanted to acquire Al Noor, a UK-listed, Abu-Dhabi-based hospital operator. However, it was unable to get regulatory approval for the deal.

> Slaughter and May solved the issue by conducting a reverse acquisition, whereby Al Noor actually ended up acquiring Mediclinic, a company eight times its size. Lawyers arranged a complex bridge-financing deal and a unique tender offer, which allowed Al Noor to make the purchase. Mediclinic is now a FTSE 100 company and the largest healthcare provider in the Middle East.
>
> This case demonstrates how legal M&A is about getting things done. In this example, the focus was on creating value by merging Mediclinic and Al Noor. Deal done. M&A is hands-on deal-making, and every corporate should align itself with the very best law expertise that their money can buy.

"I was once involved in a deal where a corporate was attempting to buy a company from a private equity owner," says Peter. "Our team was supported by one of Europe's leading law firms but still, the negotiations went slowly as we had not foreseen the seller's need for limiting its liability on management's representations and warranties."

"Usually in these sorts of deals, the lawyers would create an escrow account in which the seller would leave 10–20% of the enterprise value as collateral. However, this private equity seller needed a closed-end deal in which the buyer could not freeze part of the total payment within 2 years of the deal. The solution was to take out M&A insurance, which actually gets the seller off the hook while the buyer still has a guarantee for deal issues such as product liability, tax, and so forth, even after the 2-year period has expired. In the end, the deal was delayed by two weeks while the M&A insurance was put in place."

The survival period of liabilities on various representations and warranties is usually 5 to 7 years. However, negotiations stalled as the (outstanding) legal firm representing the buyer needed a time-out to discuss M&A insurance. The seller's law firm had done several private-equity deals on both sides, so they were experts in the field. Even with the world's leading law firms, a lack of knowledge (for this particular legal team) prevailed. However, the law firm was able to arrange knowledge overseas in their own firm, and the deal was back on track within 24 hours. You may sometimes require highly specified insight when it comes to complex matters of M&A. That is why you should never ever be cheap on legal advice.

There is one issue with M&A legal advisors which has the potential to cause conflict. The largest legal advisors get a lot of referrals from investment banks. They are not in the pocket of these investment banks at all, but if they have been connected to one of your M&A deals because the investment bank made the connection, there could be a problem. If your legal advisor is uncertain about a deal which was recommended by the bank, and ends up in a 50/50 dilemma, she may be influenced by her past and future relationship with the investment bank. Since there is a 50/50 chance of success anyway, she might push the deal through so that the bank can secure its 'success fee.'

That is why 'taking ownership' is such an important part of the M&A Formula. The CFA will dictate the terms of your relationship with all your external advisors, and it will protect you every step of the way.

In various negotiations, we often find that leading law firms are on board with the M&A CFA. Still, not one of these leading law firms would dare to recommend the framework to their corporate clients, even though these clients would greatly benefit from an M&A CFA. Why is that? Two words: investment banks.

Most, if not all, large corporates and SMEs are better off supplying their internal team with external M&A-focused lawyers. But just make sure that they have hired according to the M&A CFA.

What Can a Company Do by Itself in Terms of Legal M&A? Once a project has gone through a deep dive in relation to the business model fit, the target must be acquired and the whole organization must act decisively and with high speed. In the earlier example, a deal was delayed by two weeks because the buyer had to put M&A insurance in place as per the seller's wishes. The purpose of legal and compliance in the M&A Launchpad is to try and foresee these unexpected occurrences and prepare accordingly.

When you are putting in place your in-house legal team, the first step is to carefully consider having access to external M&A specialists. Remember, your corporate counsel is still the boss when it comes to external legal advisors, so you must clear everything by him and be sure to define your M&A buy-side goals.

M&A Love Letters

Many companies are desperate to sign exclusivity deals with the seller and will pop the champagne as soon as this is agreed. I normally advise them not to celebrate too quickly, because exclusivity is normally just a gentleman's agreement where no break-up fee has been arranged. A gentleman's agreement is not an agreement that can be enforced in any court, and there is a good chance that at least one of the two parties will turn out to not be a gentleman at all. Even the most attractive exclusivity agreement will mean nothing if a higher bidder comes along.

What is much more important is the LOI, or 'M&A love letter.' This is where all the details of the deal are laid out in black and white. In some cases, the seller may ask for an initial valuation, particularly if the buyer insists on exclusivity. This may take the form of a non-compete clause (for instance, "we will not enter this particular country in the coming 12–14 months if we get access to your company data and decide not to buy your company anyhow"). Non-solicitation could also be required from the seller (for instance, "if you talk to my best people and start building a relationship with them, you cannot hire them within the next two years").

A company needs to discuss these things internally in its M&A Deal Committee—a discussion that needs to be led by the corporate counsel and preferably assisted by an external legal advisor. Spend the time and money on this, because it builds up your M&A Launchpad and you don't want to miss an M&A target. Becoming a successful corporate acquirer demands at least as much pre-match study as when Serena Williams is planning her top matches. Too many companies simply do not prepare themselves enough.

In cross-border deals, it is crucial to be aware of the local laws. You also have to consider cost. Too many European and Asian firms choose to follow US laws, even when this isn't beneficial for them. Firstly, the USA is probably the most expensive place to litigate on the planet, and secondly, these European and Asian teams have not necessarily been trained in US law, and they may end up arguing against M&A Wall Street lawyers—the highest-paid external legal advisors on earth.

Your legal terms should also be pre-agreed in the M&A Launchpad when it comes to alertness to any specific issues which could cause havoc (e.g. pension liabilities). What kind of information do you definitely require in the data room before commencing the deal? In some jurisdictions, you have to conduct integrity due diligence (IDD). For instance, food and beverage companies will be looking for access to water and food safety standards. Most companies will also be asking for in-depth research on sustainability and corporate social responsibility issues such as child labor, or long-term issues with product health impact. Define what is important for you and your company.

It is also worth considering how to actually acquire the firm. Two different methodologies are the 'locked box' and 'classical completion' accounts.

There are many pros and cons for these two principals, but the 'locked box' mechanism is increasingly taking over as the preferred purchase price mechanism. In practice, there will often be a discussion over net working capital and net debt (i.e. the difference between realization and estimation). The locked box approach is simpler, as it doesn't tie up both parties. The price is certain, but the seller will not get any benefit from continued operations in the interim period. The buyer needs to debate debt and working capital at this early stage and increase reliance on warranties, which normally requires an enhanced due diligence process.

The 'classical completion' method may speed up negotiations but it also offers more potential for dispute between the buyer and seller. The seller bears all the risk of the business, right up to completion.

It is down to you to choose the option that works best for your company. If the seller insists on one model and you have a preference for the other model, you need to work out—ahead of time—how you might find a middle ground.

Arranging a company-specific M&A buy-side legal framework is highly recommended for any company that wants to act quickly and decisively in corporate M&A. All these pre-made arrangements can be made internally at any time, and will allow you to focus on the M&A deal at hand instead of learning needs and wants from your company during the deal process.

Of course, these are not the only pre-made arrangements a company should put in place when working to become great at M&A.

Funding and Corporate Finance

Speed and ease of execution are very often key in M&A. If you are not one of the very few large corporates who do M&A deals entirely by yourself, you may be used to a lengthy process during the M&A 'beauty parade' which goes something like this:

- Your company will identify a series of appropriate M&A advisors.
- They will come and pitch to you.
- You choose one or two.
- You negotiate an M&A engagement letter.
- You spend several weeks in endless discussions on the non-disclosure agreement, intellectual property rights, production matters, and so on …

Before you know it, two or three weeks have passed and you are still at the starting line, while your competitors are miles ahead in their discussions.

Corporates sometimes waste the important first few weeks of the race in the pit. But it is easy to avoid these issues.

Instead of going through the arduous process listed above, savvy corporates follow a non-traditional route which has a proven success rate.

Firstly, they design their own M&A engagement letter, which has been fully customized to match their own needs and wants. Think about it—an average consumer today often wants something that's unique to them, and when the future of your company is at stake, why would you settle for anything less than perfect? Yet still many corporates will sign legacy M&A engagement letters instead of customizing their own.

That is totally outdated. Not only does it result in painful onboarding times, but it also means an insurmountable pile of legal risk that can easily be avoided. Not to mention tail fees and monthly retainers, which are particular issues when companies are looking at lots of transactions at one time. Forget what the investment bankers want, the only way to get what *you* want is to create your own company-specific M&A engagement letter.

Secondly, set up your own M&A panel and choose three to five banks which fit with your M&A Launchpad. Make sure that these banks can agree with the terms of your own M&A engagement letter upfront and can support your company financially if need be. This way, you can reduce your search to a matter of hours rather than weeks.

Thirdly, in order to support fast onboarding, appoint an M&A Deal Committee within your own organization. Use an embedded M&A RFP in your own M&A CFA to align all data from your fixed M&A panel, and then the M&A Deal Committee can appoint the investment bank in minutes. You don't even have to meet up—it can all be handled online or via telephone.

The same process can be followed when it comes to your Non-Disclosure Agreements (NDAs)—never again will you pester your Legal department with such a boring issue.

All you have to do is create your own M&A engagement letter using an M&A CFA with a predefined NDA, RFP, and M&A engagement letter as the three pillars. It's as simple as that.

Q&A with Klaas Springer, Director of Corporate Treasury at FrieslandCampina

Q: Why does your company do M&A and what is the Treasury's role in M&A?

A: The most important job for the Treasury function is to secure the funds for the running of the business, as well as funding new M&A activity. We keep those funding programs completely separate in order to ensure that M&A does not take away funding from the running of the business.

Still, securing funds for M&A is something we take very seriously. M&A enables our overall corporate strategy in transforming the company and accelerating the future revenue growth that we need.

We co-operate very closely with banks and investors when it comes to our M&A activities in order to secure long-term funding. We share our expected average takeover capacity every year to make sure both investors and banks are onboard. In the past few years, we have included a new type of financing source—Multi Development Banks (MDB). MDB funding differs from traditional funding as it provides us with a longer-term commitment than we would normally get from banks. Besides, there is risk participation from the MDBs in the form of equity capital.

Last year, for instance, we acquired 51% of the Pakistani company Engro Foods. We worked closely with MDBs on this transaction and closed long-term financing agreements with the World Bank and the Dutch Development Bank. On top of the long-term commitment and equity participation, which mitigates our total risk in the project, we now also have a close allied partner (the World Bank), which could be helpful if things are not developing as we hoped.

With regards to diversifying our funding sources, we have also issued our first Green Bond (a 'Schuldschein'). Our company is a great believer in sustainability and we try wherever possible to show we are green. It is also something we look for in M&A transactions.

Q: You have an M&A Deal Committee—what is the role of the Treasury?

A: The fact that the Treasury department is a member of the M&A Deal Committee makes it possible for us to support M&A and prepare the company and all its financing partners (banks, investors, MDBs). It is absolutely essential for a company with growth ambitions in M&A to be early in the loop when it comes to Treasury, in order to act early and be prepared with the financing and risk management tools when the M&A deal starts to unfold and everything gets busy. A lot of time is needed, particularly when you are involving bond investors and MDBs.

Needless to say, we also select the financing sources for M&A transactions as well as the investment banks/financial advisors. We have almost 5 years' experience with M&A CFAs and I must say that it has saved our company a lot of trouble—we can now spend all our time finding the right financing for our M&A projects and devote as many resources as possible to each and every M&A project instead of, as previously, having endless discussions with the banks on M&A engagement letters—particularly fees and other Ts&Cs. Whenever a new contender bank comes knocking on our door we do consider them, but always with our own M&A documentation. In that way, every bank is also guaranteed the same treatment. We simply do not sign M&A engagement letters from investment banks, but they are all treated 100% equally.

We have a balanced portfolio of a few external M&A advisors, of which a couple are domestic banks. They are not in the global M&A league tables alongside Goldman Sachs, Morgan Stanley, or JP Morgan, but they have certainly earned their right to pitch for our M&A business as these banks have provided our company with unusually long-term M&A bridge financing. We find that some reciprocity towards domestic banks is highly justifiable, as they also guarantee funding for M&A projects—something that does not come from the investment banks, who often refrain from direct lending activities. We have to look at the full packet and you don't always fully benefit from choosing the highest-ranked investment bank in a particular region.

What we have noticed from all banks, be it investment banks or domestic banks, is that they are all highly interested in M&A assignments. Not only do they get to earn their fees, they also take great interest in the 'customer intimacy' opportunity that follows an M&A engagement. The banks will meet our CEO, CFO, and other key people from our organization. You really become part of our inner circle in those M&A jobs as a banker, and it's hard to really come any closer to our company as external.

5 Accelerating Your M&A Formula

Digitizing M&A

Anything can be digitized, even M&A.

Corporate M&A is one of the last remaining sectors which is stuck in analog mode, but this is changing.

As you will know by now, the authors behind this book have an optimization curse in life that forces them to keep looking for ways to improve any corporate process. The first step was to compare the in-sourcing of financial services with the procurement of A4 paper. Now we would like to challenge your existing M&A processes by showing you that it is possible to digitize any industry in order to improve efficiency.

The worlds of transportation, banking, public services, healthcare, and retail are a few obvious examples of digitization success. So we challenged our team to come up with a business that could never be digitized. The winning suggestion was 'waste management.'

Surely there was no way to digitize an industry that relied purely on manual labor and a physical product (trash). We all see the garbage trucks driving around our neighborhood; men jump out, grab the trash cans, and tip them into the back of the truck. Not a computer in sight.

We were wrong. In fact, one waste management company has been digitized for a very long time, and it has managed to make resource savings of up to 33% as a result.

San Diego waste management firm Daily Disposal was one of the first of its kind to see the efficiencies of digitization. By equipping each truck with a tablet, they were able to track each and every trash collection, and record any issues en route (for instance, trash not being left out on time, inappropriate items left for collection). This allowed the firm to figure out the 'go' and 'no go' spots, so staff were not wasting time collecting rubbish from non-paying customers, and could therefore focus more resources on their paying clientele. This reduced the pick-up time on every truck by an hour a day, resulting in a saving on fuel costs and increased (paying) customer satisfaction.

If a waste management firm can do this, so can M&A.

Strengthening the Formula

All good M&A ideas are born out of the belief that there is a competitive edge which can be obtained from joining with another firm, rather than going about things on your own. It is the belief that you can do great things together.

As firms embrace agility at all levels, ranging from production to management, digital companies also use agility to secure continuous innovation across their product portfolio. The ability to innovate becomes an attractive competitive parameter itself, so digital readiness is a no-brainer. Functional M&A software rocks, and we have not even seen the beginning of a paradigm shift yet.

How does the M&A Formula become stronger through digitization and the use of functional M&A software?

Business Model-Driven M&A

When you digitize, you force your organization to codify its processes. We searched for an IT platform that was 'plug-and-play,' allowing you to build your business model drivers directly into it. This means that everyone in your company will automatically be aligned with your business model drivers, and you can move forward with greater speed and accuracy.

Every company and each M&A project has a different set of objectives and sources of synergy (e.g. costs, margin, talent, revenue). Defining these processes and systemizing your workflows does not dampen your viewpoint, but allows you to set your more granular objectives by using the Goldman Gates.

Leadership and Communication

The team with the best people wins. By digitizing your M&A, you open up the possibility of giving everyone an equal insight into your M&A drivers. This allows you to spread your message instantaneously across the whole organization, ensuring that everyone can say: "I know why we do M&A and I know what is expected from me."

Imagine watching two infants playing together. As a parent, you always encourage your child to share their toys. Typically, they are reluctant, especially if it is their favorite toy. But as the parent, you want to encourage them to share—or as the grown-ups would say, to 'collaborate.' Sometimes companies need to have the confidence to share their favorite toys. Get down from the M&A ivory tower and ask the real people, with the true insights into your business model fit, what they think about your next acquisition or merger.

By bringing together people from different parts of the organization, you are encouraging a culture of transparency and openness. Likewise, open architecture encourages companies to play together and share their favorite toys, without necessarily relinquishing their property rights. It allows firms to collaborate and leverage the digital successes of complementary businesses, in product and in process. Consider an M&A Deal Committee in your firm or another kind of CTP which represents the common knowledge of your firm. Avoid negative biases (selection, confirmation, cognitive biases). Voices must be heard and hidden agendas seen. The shareholders want total return, not M&A pet projects.

Taking Ownership

One of the most important arguments for digitizing is that it requires you to take over all the processes yourself and create your own M&A formula, independent of external advisors such as investment banks. You have to do your own homework, put the time in, and create something which is perfectly suited to your organization's unique goals.

You should always expect granularity and cost transparency from your service providers at an individual transactional level. This can only be achieved by having access to all available data, ranging from conventional customer metrics (such as age, income, product usage, and lifetime client value) to more behavioral data characteristics (such as customer lifestyle, decision and choice science).

The ability to process and structure all data not only leads to more objectivity and factual decision-making, but can also leverage previously under-utilized opportunities to scale data and enhance the product.

Case Insight: Midaxo
(see more on www.midaxo.com)

Process is critical in digital ecosystems. Midaxo is an interesting case of digitizing processes and mapping workflows in M&A. It has developed a cloud-based software platform designed to help companies define their M&A Playbook. The platform is used by the likes of Hewlett-Packard Enterprise, ADT, Philips, PTC, and AvidXchange. Automating, mapping workflows, and capturing all available data are the first steps toward deep digital adoption.

As we already know, preparation is everything when it comes to business model-driven M&A, and the right M&A platform can play a vital role in this process. By putting every deal through a standard set of considerations, you can reduce risk and help your organization learn from its mistakes.

If collaborative IT is akin to two infants playing together, then IT platforms such as Midaxo are tailored for big groups of children learning to play as a team. These platforms do not have room for silent players—it's actually not even possible, as everyone can see what's going on in an M&A deal at any particular time.

When you force an existing organization to move away from the bad habits of old-style analog M&A and into the M&A Formula, the first thing you need to do is make it business model-driven. This must be built into a support system like Midaxo. You will also have to create a CTP environment as opposed to old-style emails, PowerPoint presentations, and management meetings. A system like Midaxo supports this and works well with an M&A Deal Committee to cascade good and bad news throughout the organization. The real advantage of CTP platforms like Midaxo is that they break down silos and therefore increase the combined mutual intellectual knowledge. Who doesn't remember the Emperor's new clothes?

In the absence of this technology, M&A process definition initiatives often fail to become actionable and reach their practical value. In fact, in many cases the long process documents produced in these initiatives end up collecting dust. If this data was presented in a clear, uniform manner, it would be quick

(continued)

(continued)

and easy to pull out the key points and learn from the costly mistakes of past processes.

Technology makes it possible to define, realize, and improve the process in a practical way—and the process is what creates the value.

The Midaxo M&A Platform

The key features of the platform aim to establish processes for a template library, deal pipeline management, management of individual deals, collaboration, and reporting. They enable the M&A team (including the customer's employees, acquired employees, consultants, advisors, etc.) to systematize and manage their complex activities, from DIY deal sourcing and valuation through to the transaction and integration processes. The platform becomes the interface to planning, executing, and reporting on the M&A process and each individual deal.

The project manager can define and communicate clear project plans, based on best-practice templates. They can work collaboratively with all stakeholders to execute the plans efficiently and in a disciplined way.

One of Midaxo's design principles was to enable M&A team members to focus on the work rather than spending time on learning a complex tool. Providing them with clear advice and instructions at the level of their tasks lowers anxiety, expedites the process, minimizes mistakes and risk, and drives positive experiences with the project.

The big criticism of digitization is that it removes the 'human' element from the process, and changes the level of service offered. Wrong assumption. More myth killing. We just experienced how you can change your organization to a network-based way of doing things instead of a hierarchical system. That is *about* the human element—having people work much more closely together. We produced this book in Google Docs, for instance. This was our CTP in order for co-authors, proofreaders, Wiley, professors, and the M&A panel to work together in a cohesive way, avoiding emails and Word documents with no back up, allowing everyone to contribute at any point in time. Albeit we did sometimes have various rights with regard to read-and-write, which is what Midaxo offers your team also.

Creating CTP for your people working with M&A is not even the beginning of the human element. Wake up—you are in a digitized age. You can arrange meetings on the fly, hold conference calls remotely, share viral video clips, and automate all your admin processes. It is even possible to test your own bespoke M&A Formula through gamification (see more on www.fixcorp.co). The use of flight simulators has a variety of reasons—a "first time" pilot is self-explanatory perhaps but even highly experienced pilots do training and work with manufacturers on design and development of the aircrafts. An M&A simulator, testing your M&A Formula, can do the same for your organization no matter if you are an M&A savvy corporate or a first time SME acquirer. Of course you can digitize your M&A business.

The M&A Dashboard

A lot of companies can benefit even further from the incredible human factor in M&A. But the second point in our M&A Formula is 'leadership' for a reason—if you don't understand the importance of the human side of M&A, you will never be able to leverage it like our M&A Elite. HR is usually seen as a 'soft factor' in M&A, but if you want to play hardball and make M&A successful, you also have to use this soft factor. Remember, all over the world there are people fighting each other many years after one group was 'acquired' by another through colonialization, annexing a piece of land, closing down social media, or limiting voting power to create a dictatorship. This is the oldest story on earth, and it has caused problems that persist for hundreds (or even thousands) of years. The acquirer cannot rule properly as they have to deal with dissent, and the acquired people will never feel a sense of belonging. If you can't create an atmosphere of collaboration from the very beginning, then 'acquired' people will fight you.

Again and again we have tried to emphasize the importance of 'leadership' in the M&A Formula. Weak leaders constantly underestimate the value and importance of their staff, and this will always backfire spectacularly when M&A activity takes place.

Prior to, during, and after the whole M&A process, employees are fraught with severe needs and problems, and a lack of awareness of your employees' concerns can result in M&A failure instead of success.

We recommend creating an M&A Dashboard to complement your M&A Formula. This Dashboard will help you to stay focused on your 'leadership' responsibilities, by involving and informing staff every step of the way. You may think that you have this covered, but it is easy to lose sight of your staff's happiness when you are busy focusing on business model drivers and drawing up CFAs.

"The M&A Dashboard is like the dashboard of a car," says Peter. "You know what's there, but you wouldn't want to drive without it. Of course, I decided to test this theory so I got into my car and covered up the entire dashboard, then started driving. The first thing that struck me was that it was apparently pretty easy to drive the car without a dashboard, although I did feel a bit uneasy when going through town as I didn't know my speed. I then considered the long-term repercussions of driving 'blind.' I forgot to check fuel and even after a short while I started to wonder whether I was running out of fuel—I genuinely had no idea."

When you cover up your dashboard, you take a lot of things for granted. It's the same with M&A (see Table 5.1).

Table 5.1 M&A Dashboard

Problems During M&A Processes	M&A Dashboard Benefits
Soft facts remain undetected on both sides (buyer and target). Employees are in the constant-stress field of ignorance and uncertainty.	With an M&A Dashboard, you always know about your employees' moods and specific concerns. M&A integration leaders can "systematically plan interventions to smooth the human integration process."
Relevant M&A information is filtered by and limited to key positions. Individual employees' needs and ideas are not known.	Every individually relevant employee is involved in the process and gets a voice.
Classical questionnaires or personal interviews show only snapshots without enabling managers and employees to implement any measures/actions.	An M&A Dashboard uses intelligent and self-learning pulse surveys.
Business model drivers can be communicated, but is the information also understood by the employees?	Direct communication of "Why we do this deal" and control of business model drivers and their metrics, actions, and/or verification of communication of what is expected from *you* in this deal.
Hazard grows and lurks beneath the surface. M&A can lead to employee resistance.	Timely intervention before the grapevine takes over.
When executives finally learn the true story about the well-being and worries of their employees, it is usually too late to react …	This way you always have an ear to the ground, can intervene and initiate countermeasures immediately.

Digitizing: The Human Side of M&A

During M&A processes, employees are confronted with many uncertainties. What will happen to me and my position in the company? What will happen to my department? What changes are coming? How will these new processes impact on our customers?

In order to better understand the human side of M&A, let's have a closer look at two special concepts: Anxiety Theory (Adams) and the Organizational Justice Theory (Greenberg).

Employees are exposed to these two theories throughout the different phases of M&A integration (see Figure 5.1).

Anxiety Theory (Adams)

It is a general observation that employees experience a high degree of anxiety when facing the possible occurrence of M&A. Prior to the actual combination of acquirer and target organization, employees have fears about the consequences for their future jobs and careers—mostly projecting worst-case scenarios.

To counteract anxiety-related stress, managers are advised to apply formal communication measures to provide employees with clear and

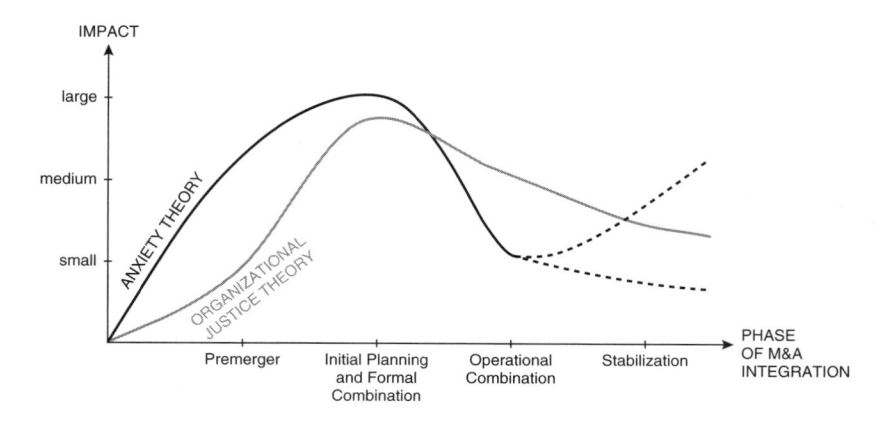

Figure 5.1 Business theories applied to M&A

concrete information about expected consequences for the organization and employees' jobs.

Organizational Justice Theory (Greenberg)

Employee reactions to an organizational change such as M&A can be influenced by the following three types of fairness perceptions:

- Distributive justice, which is the fairness of sharing rewards and costs among organizational members relative to an individual's standard of fairness.
- Procedural justice, which is the fairness and transparency of decision-making processes.
- Interactional justice, which is the extent to which organizational members are treated with politeness, dignity, and respect.

When employees see themselves as being treated fairly, they are more likely to develop attitudes and behaviors in support of change, even under conditions of adversity and loss.

If you don't have an eye and an ear on the human side of M&A processes, then your organization is at risk of quickly experiencing psychological and physical illness, resulting in employee turnover and counterproductive work behavior.

Now, Science Meets Business

On the basis of organizational behavior theory, you can monitor your M&A Formula in real time. SMA Research Lab created two axes for monitoring the M&A Formula: from today to the future; and from the individual employee to the entire organization. The software is known as Business Beat (more on www.business-beat.com).

As you can see from Figure 5.2, this results in four fields with specific management goals, ranging from securing commitment (today's

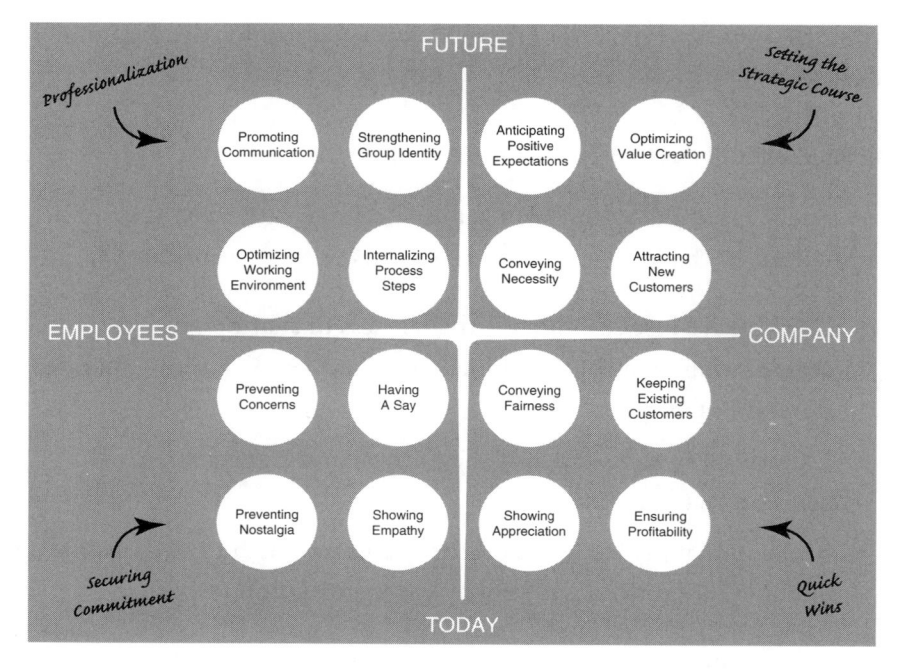

Figure 5.2 Monitoring the focus and progress of your M&A

achievement on an individual level), to generating quick wins (today's achievement on an organizational level), to accomplishing professionalization (tomorrow's achievement on an individual level), and setting the strategic course (tomorrow's achievement on an organizational level). Each field includes very specific challenges to be tackled by M&A integration leaders, as well as management, in achieving the change-related goals.

Digitizing M&A with Business Beat

Business Beat is a functional M&A software that allows you to take care of your employees in real time, and make them feel that they belong. It does this by sharing information with staff regularly and thoroughly. There is no such thing in corporate M&A as over-informing. No-one has ever complained that they have been given too much information during an M&A transaction.

Based on business research, a special question pool for M&A processes was developed by Business Beat in collaboration with the SMA (Strategy, Mergers & Acquisitions) Research Lab: instead of a delayed, annual survey about job satisfaction in general, employees were confronted with particular M&A challenges at very short intervals. Also, employees were always directly involved in the processes and allowed to contribute to a more effective and efficient M&A integration process. And for the first time, they had the opportunity to experience developments as fair.

Part II
The M&A Formula Applied

Part II

6 Case Insights (CI) and Research Insights (RI)

CI1: DSV: From Ten Trucks to €10bn

The world is not exactly short on third-party logistics (3PL) firms. It is a cut-throat business with low entry barriers and a very simple objective—bringing stuff from A to B, delivered to the right place on time, not damaged, as quickly and cheaply as possible.

Many firms buy services from 3PLs as part of their distribution models, and the best-known firms (DHL, UPS, Kuehne & Nagel, Schenker, Nippon Express, Sinotrans, C. H. Robinson) are some of the most sought-after companies in terms of M&A. Yet a Danish company called DSV has risen to the top of the ranks to become one of the most successful 3PL firms in the world, with an M&A track record so strong that it has gone from ten trucks to €10bn since its creation in the late 1970s.

The global transport and logistics business operates with razor-thin margins, and only a few companies can reach the top. As a general rule, the top 10 3PL companies come from the top 10 countries by population and by export—after all, the more people there are, the more goods there are to move. It is not surprising, therefore, to see US, Chinese, and Japanese firms rise to the top of the 3PL league. But what is a Danish company doing there?

DSV was established in Denmark in 1976 by nine independent truck drivers and a business developer called Leif Tullberg. Denmark was, and still is, hovering around number 36 in terms of global exports, and 111th when it comes to population. The disproportionate export ranking is mainly due to the fact that the country holds a very high export ratio of 53%, whereas most other countries have export ratios of about 10–25% of their GDP (with the exception of Germany, which is ca. 43%). It seems as if you cannot become top 10 in global transport and logistics if your company is domiciled in a small country.

Yet, DSV has risen to become a solid top 10 global 3PL business. A global leading 3PL firm is normally a combination of being born in one of the world's most populous countries with the highest exports in the world. Despite coming from 'Lilliput,' DSV did it anyway (see Table 6.1).

Table 6.1 CI1: DSV operational excellence with cost as business model driver

10-year return on DSV stock	
Initial investment 01/01/2007	€100
Would have grown to	€318.97
Yearly return	12.3%
10-year benchmark return Eurostoxx	
Initial investment 01/01/2007	€100
Would have grown to	€121.09
Yearly return	1.9%

So, What Made This Possible?

In short, DSV's success is down to its relentless focus on business model-driven M&A. This is a company that drives hard on reaping high cost synergies for each corporate takeover and at the same time 'connecting more dots' on the world map, thereby gaining access to more global clients with a need for distribution and logistics in the air, at sea, and on the road.

In the past 10 years, like the other Global M&A Elite companies, DSV has significantly outperformed the general Eurostoxx. In this same period, DSV has grown about 30–35% organically and 65–70% purely by acquiring other 3PL firms globally.

It is perhaps the most obvious evidence yet that corporate M&A can create value for shareholders. While other companies in the M&A Elite have provided shareholders with an almost similar or slightly better total return on their shareholdings, they have not grown quite as much as DSV in relation to M&A, and they have arguably experienced a slightly more lucrative business environment than the transportation sector.

Despite this, DSV has used corporate M&A to become a strong member of the M&A Elite, and one of the top 3PL companies in the world.

In 2016, DSV completed the acquisition of the US company UTI Worldwide. This was a typical deal for DSV; UTI was a company that was performing below best in class with an operating cost which was much too high.

The number one business model driver for the acquisition was cost synergy (see Figure 6.1). DSV aimed to reap synergies of about $225m, or approximately 5–6% of revenues. While this may seem low compared with other members of the M&A Elite (e.g. RB, which regularly hits 12–14% in cost/sales synergies), this is only because DSV is a service company, and does not sell products. In the service world, 5–6% is a very high figure indeed.

DSV was able to find about half of these cost synergies within the Finance and IT departments, and the rest came by closing down duplicate headquarters, distribution centers, and so forth. DSV carried out an IT conversion within 12 months of the acquisition, and moved all of UTI's volumes into DSV's existing systems. There was also a 30–40% reduction in white-collar workers, which added to the total cost synergies.

In the end, the UTI deal added another 179 days of revenue to DSV, or 46% of its revenue.

This was a very typical M&A deal for DSV. What is particularly remarkable about DSV is its ability to maintain focus on its M&A business model drivers. It has been best in class for operational excellence in 3PL for years, but it knows that it won't add shareholder value by buying a company with similar trading multiples and paying a 20–30% premium for the privilege.

In the past, DSV has successfully integrated ABX and Frans Maas into its business (these two companies were about 40% and 60% of DSV's own size). It is unusual to see a company buying so many relatively large competitors, but DSV is no stranger to buying considerably bigger companies.

In 2000, DVS bought DFDS Dan Transport (a 3PL company that was five times bigger than itself) and in 1997, it bought a company called Samson, which was twice the size of DSV at that time.

DSV shares many similarities with other M&A Elite companies. Like Danaher, the top management has many business reviews based on input from empowered back-office functions, in particular finance. And like ASSA ABLOY, local management around the world has been incentivized via bonus schemes and sometimes stock options in DSV. This means that all members of the top management have their noses in the same direction, and every one of them is motivated to create results for DSV worldwide.

What is unique about DSV is its ability to pull off relatively heavy acquisitions. Not only is the world of M&A known for an average failure of at least 50%, but the idea of acquiring firms 50–500% of your own size is for in most businesses.

But what DSV offers is the opportunity to add value to these larger companies, by getting rid of bad management and introducing the highest operational excellence.

The DFDS deal actually came about from two of DSV's major shareholders, who suggested: "Why don't you buy this company, integrate it, fire the bad management, and run it like DSV?"

The Frans Maas deal came about in a similar way. DSV's (then) CEO Kurt Larsen was contacted by an investment bank representing two of the largest shareholders in Frans Maas. They were not happy with the company's performance and felt that value could be created in a takeover by DSV. It doesn't take much of a genius to calculate what value can be created if an underperformer is taken over by a company known for its operational excellence, so long as the integration goes well.

DSV initially approached Frans Maas with a deal which involved a ratchet feature, by which DSV would pay the investors more than the promised €38 per share (at a time when shares were trading at €28). This premium equaled 35% or slightly above the average premium that companies normally pay for the control of stock-listed shares (about 30%). Mr. Larsen also suggested a merger between the two companies, but the top management turned him down. At this point, DSV revealed that they already controlled almost 50% of the shares, thanks to the intervention of the firm's investors. Checkmate.

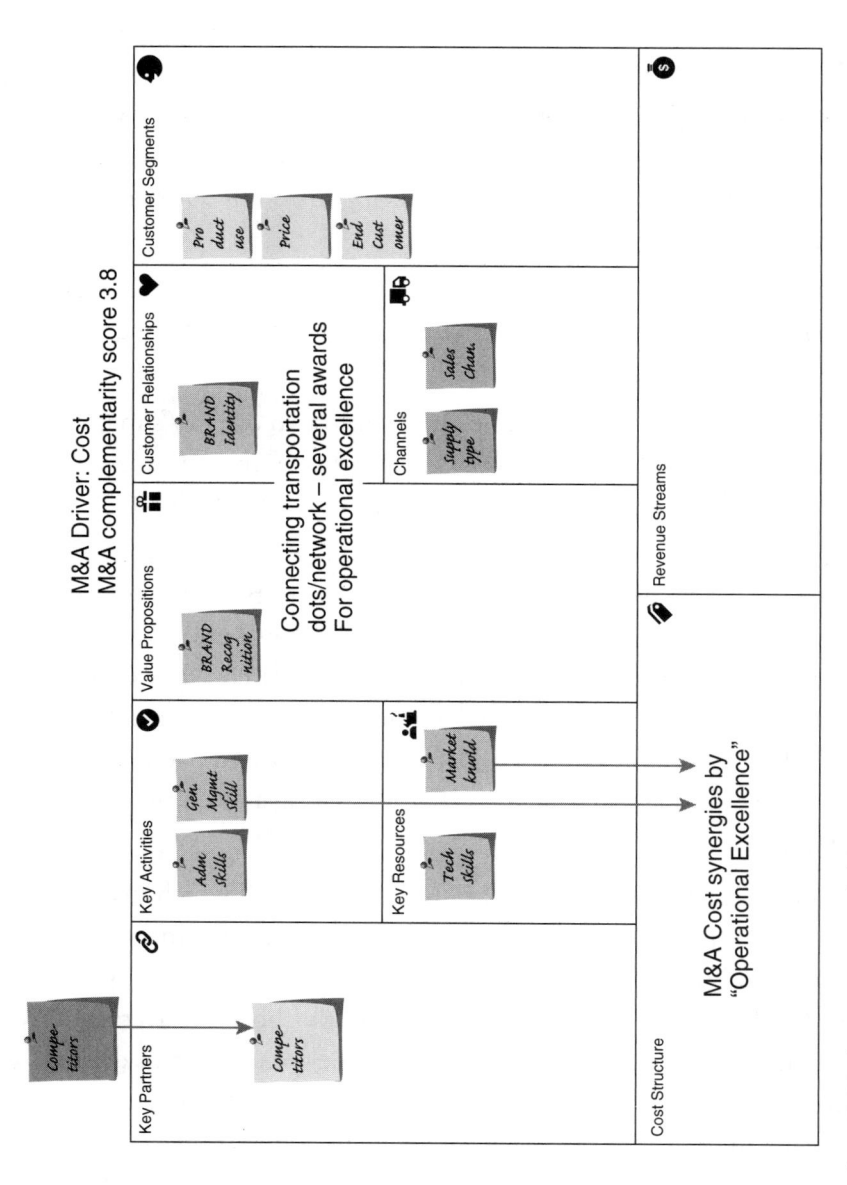

Figure 6.1 DSV business model canvas

This rather brutal approach worked, the acquisition took place, and the bad management was duly replaced.

As with all the other M&A Elite companies, it is crucial that the acquirer is best in class—how else could you prove your operational excellence?

DSV has proven its ability time and time again. It has received countless awards in various countries and for 3PL disciplines by sea, air, and road. DSV has also won three major awards for best annual report by leading equity analysts, who find the company's reporting highly valuable and insightful—and when externals are impressed, you must be doing something right! Even the company's truck drivers have received awards for excellent driving and safety measures.

For DSV, this operational reputation allows it to focus on its key M&A business model driver—cost synergies (see Table 6.2).

Table 6.2 RI1: DSV—from ten trucks to €10bn

Research Insight title	Cost as a business model driver
Case Insight to discuss	DSV
Context from Peter Secher	Cost as a business model driver based on operational excellence—M&A complementarity total score 3.8
Research Insight content	**Point:**
	The transfer of complementing resources allows firms to reduce costs and 'plug in' the acquirers' processes and profit formula.
	Counterpoint:
	If the acquirer has similar resources, a 'plug-in' might not be possible and the cost structure would just be replicated. Further, if the business model depends on key employees, disruption should be minimized.

(continued)

Table 6.2 *(continued)*

	Contingency: Speed in integration matters. Without the transfer and sharing of resources and capabilities, no value occurs out of the acquisition.
References	**Point:** King, D.R., Slotegraf, R.J., & Kesner, I. (2008) Performance implications of firm resource interactions in the acquisition of R&D intensive firms. *Organization Science*, 19(2): 327–340. **Counterpoint:** Christensen, C.M., Alton, R., Rising, C., & Waldeck, A. (2011) The new M&A playbook. *Harvard Business Review*, March: 49–57. Schweizer, L. (2005) Organizational integration of acquired biotechnology companies into pharmaceutical companies: The need for a hybrid approach. *Academy of Management Journal*, 48(6): 1051–1074. **Contingency:** Bauer, F., King, D.R., & Matzler, K. (2016) Speed of acquisition integration: Separating the role of human and task integration. *Scandinavian Management Journal*, 32: 150–165.

CI2: RB: 'King of OTC' Improved Value Proposition and Client Relation as Business Model Driver in M&A

Pain or gain? Understanding value propositions and customer relationships.

When my daughter was sick recently, she asked me to buy her a packet of Strepsils. Not lozenges, or boiled sweets, or prescription medicine—she only wanted Strepsils. On my way to the pharmacy I thought back to my own days as a boy on my parents' farm, when I too reached for the Strepsils at the first sign of a sore throat.

In fact, 'Strepsils' have been a household name since the 1950s, everywhere from the UK to Canada to rural Denmark. It is an incredibly well-known brand across the world, despite the fact that it has never advertised its products any more than other brands.

The Strepsils brand differentiates itself because it has an extremely defined value proposition. According to Alexander Osterwalder's book *Value Proposition Design*,[1] in order to determine value proposition, you have to ask just one question: "What is this product or service doing in your business model?"

Osterwalder believes that there are only two correct answers to this question—the product or service must be either a pain reliever or a gain creator.

In every way, the Strepsils brand is a pain reliever. It is a pain reliever for its unwell users, and it was a pain reliever for Reckitt Benckiser (RB) when they acquired it in 2005.

If you look at the other brands in RB's portfolio, it is clear to see why they chose to acquire Strepsils. With a couple of exceptions, the firm's current brand portfolio features a series of well-known products with a distinct functionality, including a number of pain relievers and other household names which inspire strong brand loyalty.

These include: Aerogard, Air Borne, Air Wick, Amphyl, Bonjela, Brasso, Calgon, Cêpacol, Clearasil, Cillit Bang, d-CON, Dettol,

Durex, Finish, Frank's RedHot, French's, Gaviscon, Glass Plus, Harpic, K-Y, Lemsip, Lysol, Mortein, Mr. Sheen, Mucinex, Nurofen, Sani Flush, Scholl, Strepsils, Vanish, Veet, and Woolite.

It's amazing what 35 well-known consumer brands can do for a company, and how easy it must be to market and sell these products when they share a strong and connected value proposition. RB's understanding of value proposition and the customer relationship may go some way towards explaining why the company has delivered a total return of 11.30% for the past 10 years—the second highest return of our M&A Elite (see Table 6.3).

In its own words, RB is focused on organic growth and only reaches out when M&A opportunities could play an important role. While this statement may seem a little rote, it is actually a rather intelligent approach as it keeps the focus on running the business, and that is exactly what business model-driven M&A is about.

The value created by business model-driven M&A is to drive hard on the right issues; success is reached by the business model fit (see Figure 6.2). Corporate M&A must never be the main driver for a company. Investors in RB hold their shares because the company is in the world elite of corporate M&A, with an incredible ability to make value-accretive corporate acquisitions, thereby increasing the returns to shareholders.

Table 6.3 CI2: RB (formerly known as Reckitt Benckiser)

10-year return on RB stock	
Initial investment 01/01/2007	£100
Would have grown to	£396.70
Yearly return	14.8%
10-year benchmark return Eurostoxx	
Initial investment 01/01/2007	€100
Would have grown to	€121.09
Yearly return	1.93%

Speak to any professional investor, broker, or dealer, and they will praise RB's key activities, key resources, and business model-driven corporate acquisitions. But the company is perhaps most lauded for its cost synergies, which are among the highest in the world! On average, RB has doubled margins on its acquired companies and demonstrated cost reductions to sales of above 10%.[2]

In the world of corporate M&A, this is rather impressive.

Cost Efficiency Ratios

In one of my first roles, I remember noticing a growing focus on efficiency ratios, particularly when it came to cost/income ratios. This was a simple ratio comparing the total cost of the operating department or division with the total income it created. So, for instance, if a department had a total cost of €70 and an income of €100, the C/I ratio would be 70. Let's assume that the target C/I is 50; that would require the department to cut costs by €20.

As you can imagine, no-one wants to make cuts within their team, so in order to hit that C/I efficiency ratio of 50, the only alternative is to increase the department's income.

However, there is a third way to reach the ratio without putting unnecessary pressure on your team. I suggested that we work with both numbers simultaneously, to increase our top line while also reducing cost. This is not a short-term solution, but a long-term plan which may involve a one-off cost reduction of €10 and an ongoing commitment to growing the top line by 10% every year.

This solution is less traumatic than the €20 cut, and more realistic than an immediate 20% income spike, and more importantly—it works. It is not about luck, or snap decisions. It is about knowing what success looks like in corporate M&A.

RB understands this. It has proven again and again that it is possible to deliver on both measures and create an almost unheard-of level of efficiency and M&A success.

How Does RB Increase the Top Line of Its Acquired Brands?

Most people don't realize that RB is actually less focused on the products they acquire and more interested in what job they do for their customer segment profiles.

What is particularly interesting is that their consumers rarely go for the cheaper options when they are looking for their painkillers—when consumers buy a typical RB product they are not focused on the price, but what the product will do for them.

A non-branded ibuprofen 200mg painkiller can be bought for as little as a penny a tablet, whereas a branded Nurofen pill comes in at almost five times the price. Nurofen charges even more for its 'express liquid capsules,' its 'plus' range, and its 'cold and flu' tablets (which are actually 200mg ibuprofen + phenylephrine hydrochloride). You name it, Nurofen will have it; at a higher price than any other alternative.

Similar comparisons can be made with Strepsils. However, you will still buy it because when you are in pain, you want a pain reliever right away, and you want to be certain that it will work. RB products are at least two or three times more expensive than any competitor. But people aren't rational when it comes to pain relief and healthcare.

RB is absolutely fantastic when it comes to widening the value proposition and broadening its client base by using the same OTC channels. RB has a de facto monopoly in many healthcare categories, and their products are popular despite their 'premiumization.' Through years of highly focused M&A activity, RB has been able to secure higher margins and growing revenues, while still making phenomenal cost savings. I sometimes ask myself if the "Giffen Good" theory applies here—besides applying cost synergies. The best of all (M&A) worlds perhaps?

RB: A Timeline of Acquisitions

- In October 2005, RB acquired Boots Healthcare Division, bringing Strepsils, Nurofen, and Clearasil into its fold. Before anything was finalized, RB's senior management informed their investors that they expected the deal to create cost savings of £75m, equivalent to 14% of sales.

- In January 2008, RB acquired Adams Respiratory Therapeutics, which included cough treatment brands such as Mucinex and Delsym. The dealmakers promised to more than double margins, to more than 30% from an inaugural starting point of 14%.

- In July 2010, RB acquired SSL, which included brands like Scholl footwear, Sauber, and Durex. Guidance on cost synergies-to-sales was 12%, equivalent to the cost reduction/first year of SSL revenue (ca. £100m).

Judging from the overall margin improvements in RB, it seems as though RB has substantially exceeded all expectations, with each one of these deals delivering strong value for the overall brand.

'Better Business': Business Model-Driven M&A

RB is a company that truly understands its clients better than most other companies. This is because the firm takes the time to map out each product's value proposition and create customer relationships so unique that RB can chase higher margins while also reaching out to new client segments.

Investors love RB for its ability to drive down costs and increase margins, while acquired companies can benefit from the firm's world-class R&D facilities, and customers continue to love the familiar brands that they have been using their entire lives. Everyone wins.

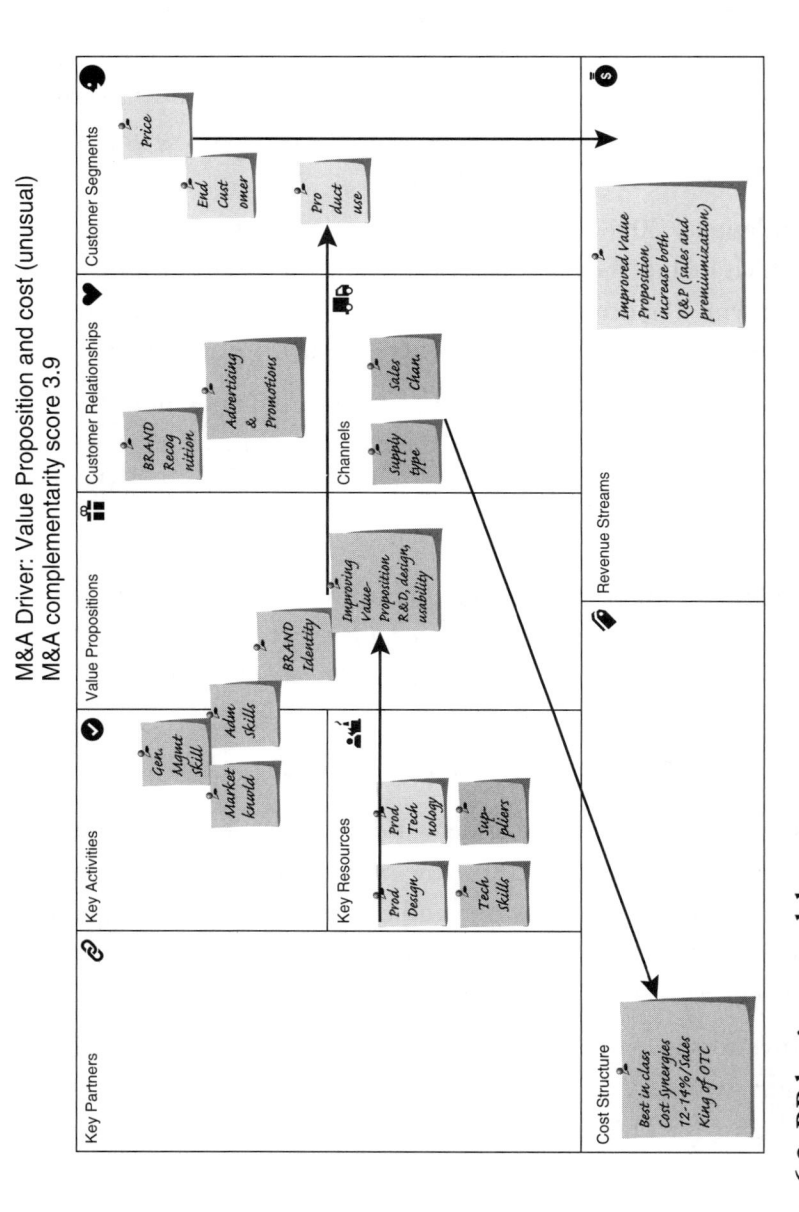

Figure 6.2 RB business model canvas

Case Insights (CI) and Research Insights (RI) 147

The business model drivers for RB deals are about two things. Firstly, a cost reduction measured to sales of 12–14% is created by acquiring products which are already being distributed via RB's existing channels. Secondly, a revitalization of the products they buy into higher gross margins and higher revenues.

Making this work is by no means easy, but by viewing the process on the Business Model Canvas illustration, it is very easy to understand and communicate to the organization that needs to deliver.

RB's executive board and the company's management team are highly regarded amongst investors, and it's no wonder. They have the same incentives as the investors, which is to create long-term shareholder value.

The company shares many other business model drivers with its peers in the M&A Elite group. A clear focus on market position, size, quality, and margins, plus strong R&D support for companies taken onboard and ongoing operational excellence in the market. This is how you become an unstoppable force in corporate M&A (see Table 6.4).

Table 6.4 RI2: Improved value propositions and stickier client relations

Research Insight title	Improved value propositions and stickier client relations
Case Insight to discuss	Reckitt Benckiser
Context from Peter Secher	Improved value propositions and stickier client relations with an M&A complementarity score of 3.9 (particularly high in product/market combination and value chain/channels)
Research Insight content	**Point:** Marketing and brand redeployment has positive effects on cost-cutting and revenues.

(continued)

Table 6.4 *(continued)*

Research Insight content	**Counterpoint:** Cannibalization effects in cases of overlapping brand portfolios might lower cash flows. **Contingency:** Organic and acquisitive growth are not a dichotomy, firms can pursue organic and acquisitive activities in parallel. Additionally, considering stakeholders improves acquisition outcomes.
References	**Point:** Capron, L., & Hulland, J. (1999) Redeployment of brands, sales forces, and general marketing management expertise following horizontal acquisitions: A resource-based view. *Journal of Marketing*, 63: 41–54. **Counterpoint:** Bahadir, S.C., Bharadwaj, S.G., & Srivastava, R.K. (2008) Financial value of brands in mergers and acquisitions: Is value in the eye of the beholder? *Journal of Marketing*, 72: 49–64. **Contingency:** Achtenhagen, L., Brunninge, O., & Melin, L. (2017) Patterns of dynamic growth in medium-sized companies: Beyond the dichotomy of organic versus acquired growth. *Long Range Planning*, 50(4): 457–471. King, D.R., & Taylor, R.W. (2012) Beyond the numbers: Seven stakeholders to consider in improving acquisition outcomes. *Graziadio Business Review*, 15(2).

CI3: Heritage Comes First at LVMH

LVMH is one of the most successful luxury brands of all time, but what exactly does it do? Sell bags or buy companies?

Actually, it does both. Firstly, as the owner of 60 international brands, LVMH sells a lot more than just bags, although fashion and accessories make up more than 50% of the profits of the group. The rest of the company profits come from wines and spirits, perfumes and cosmetics, and even watches and jewelry.

Secondly, the company does buy other companies but it has a meticulous way of running its own business which marks it out as different among our M&A Elite.

While it might be hard to pin down LVMH's 'as is' business model, it is probably even harder to predict what the next business model addition will be. Some people might think that LVMH owner Bernard Arnault is simply a collector of great brands, but the truth is that LVMH is just like any other successful serial business model-driven acquirer. LVMH is like many of the other M&A Elite companies, albeit with an understanding of heritage.

But there are perhaps two important differences in LVMH's business model-driven approach that clearly distinguish the company from other successful serial acquirers.

Like DSV, LVMH is no stranger to an unsolicited takeover—I always try not to use the phrase 'hostile takeover,' as it doesn't always convey what the shareholders think. Still, LVMH has no problem playing by its own rules when it comes to business model-driven M&A, and that's exactly how a company like LVMH should play this game. Can you create value-added growth through acquisitions? If the answer is yes, then do it. Do you need to make sure management are on board? If you don't need them, then just do it. I'm sure Mr. Arnault doesn't wear Nike trainers, but he is definitely a 'just do it' type of guy.

Perhaps the most famous example of this 'just do it' attitude was when LVMH made a play for Hermes. Mr. Arnault approached the CEO of Hermes, Patrick Thomas, to inform him of his intent to take over the firm. That may sound friendly, but it was absolutely not—prior to this conversation LVMH had built up a secret stash of Hermes shares, flouting the stock exchange rule that requires companies to flag up any position over 5%. Building up such secret positions is not legally accepted, and LVMH was fined €8m—pocket change for a company worth almost €40bn. In response, Mr. Thomas publicly slammed Mr. Arnault, making headlines across the world.

In my view, there is nothing more egoistic and damaging than a CEO publicly condemning a potential acquisition. A CEO is just working for the owner, and he or she does not represent the shareholders. Whenever there is a discussion about merging two business models, the immediate reaction should be to look for the opportunities available to both parties, not to instantly reject the possibility of a deal because you may loose your position—as this has nothing to do with the shareholders and their total return on equity.

In reality, LVMH is exactly the sort of company you would want to be acquired by. It is respectful of its brands, and the people living there seem to have a very happy working life.

LVMH has an absolutely incredible and non-intrusive way of buying these firms and maintaining the national heritage and company culture (see Figure 6.3). Walk into a Bvlgari store and it will look the same as always. Look through the window of a Marc Jacobs boutique and you will see the same clientele of young people and fashionistas. Same with the German travel bag company Rimowa, which was welcomed into the LVMH group with a "herzlich willkommen!" from Mr. Alexandre Arnault.

That, in a nutshell, is how LVMH does corporate takeovers. The new business model is to empower the existing organization and brand to build a stronger customer relationship (see Table 6.5).

Table 6.5 CI3: LVMH key resources and channels as business model drivers in M&A

10-year return on LVMH stock	
Initial investment 01/01/2007	€100
Would have grown to	€314.30
Yearly return	12.1%
10-year benchmark return Eurostoxx	
Initial investment 01/01/2007	€100
Would have grown to	€121.09
Yearly return	1.9%

LVMH: Business Model-Driven M&A

When a company is acquired by LVMH, it becomes a member of the 'LVMH Houses.' Where other companies would often re-brand or co-brand the new addition, and start cross-selling other products or services, LVMH does the exact opposite. It actually empowers the acquired organization in the value proposition and customer relationship segments, strengthening both. You will not see any LVMH branding in a Marc Jacobs store. Sometimes the people working in the store have no idea who the store's owner is. Likewise, in the mind of the customer, the LVMH brand is not even connected to their interest in Marc Jacobs. They will buy Marc Jacobs clothing or accessories because of the unique value proposition to them as a client. This is exactly what LVMH wants to achieve through its corporate M&A. However, a company acquired by LVMH will no longer have a traditional full-scale business model; it will merely focus on value proposition and its appeal to the various client segments, while simultaneously improving the customer relationship.

The acquisition of Loro Piana is perhaps the best example of how business model-driven M&A is integrated into LVMH post-takeover.

Loro Piana is a high-end 92-year-old company from Milan, Italy. The company started exploring fine fabrics just before the Second World War, and since then it has grown organically with a full-scale business model, which even includes ownership of a nature reserve in Peru to get access to fine wools.

Post-takeover, the company only needed to focus on its value proposition, customer relationships, and client segments. Antoine Arnault, son of Bernard Arnault, was appointed as chairman and he appointed a French CEO, Matthieu Brisset. Mr. Antoine Arnault stated that "what our clients want is stability and we want the Loro Piana family to still be part of this." He did this by offering the family 20% of the share capital and asking them to be a part of running the company. The Piana family agreed that there was no reason for the company to be successful by changing its strategy, and both parties pledged to maintain the quality of the brand's goods just as they had in the past.

The last time I was in Paris I saw that Loro Piana had opened a store in one of the city's best locations (Avenue Montaigne and Rue du Faubourg Saint-Honoré). What really happened was that the big and powerful global corporate LVMH made such arrangements possible for Loro Piana. LVMH has the power to get the right locations for their houses, and Loro Piana was no exception. It is a key resource not only to be able to negotiate hard for such locations, but sometimes also to be able to come up with the rental money upfront. When Asians, Russians, Americans, and Europeans visit Paris, they will see the Loro Piana brand sitting alongside other leading brands of the world.

On top of this LVMH, like any global corporate, will be able to drive down the cost of goods sold by at least 2–3%, while securing quality and delivery. Any LVMH 'house' which dreams of a big investment has access to group funding and the very best R&D. They also have access to the finest raw materials in the world, and the sheer scope of the LVMH group means that it is also possible to guarantee standards for sustainable production and the conservation of certain wildlife species.

Summary of the brands and houses within the LVMH group

Fashion & Leather Goods—16 houses, €12.8bn sales in ca. 1,500 stores
Loewe, Moynat, Louis Vuitton, Berlutti, Rimowa, Loro Piana, Fendi, Céline, Christian Dior, Emilio Pucci, Givenchy, Kenzo, Thomas Pink, Marc Jacobs, Nicholas Kirkwood, and Edun.

Wines & Spirits—21 houses, €4.8bn sales
Château d'Yquem, Dom Pérignon, Ruinart, Moët Chandon, Hennessy, Veuve Clicquot, Ardbeg, Cháteau Cheval Blanc, Glenmorangie, Krug, Mercier, Belvedere, Cloudy Bay, and 10 more wine brands.

Perfumes & Cosmetics—9 houses, €5bn sales
Guerlain, Acqua di Parma, Christian Dior, Givenchy, and some other less-known brands.

Watches & Jewelry—7 houses, €3.5bn sales in ca. 400 stores
Chaumet, Tag Heuer, Zenith, Bvlgari, Fred, Hublot, De Beers.

Other businesses—5 houses, €12bn sales in ca. 1,800 stores
Sephora, Starboard Cruise.

The reason for listing most of the brands owned by LVMH is merely to reiterate that business model-driven M&A approach. Even though it's only 70 brands (or 'houses'), you still have to concentrate now and then in order to remember, for instance, if Kenzo is really owned by LVMH. The purpose of this example is to let you envisage how hard it is for a retail customer to know whether or not their beloved brand is actually owned by the French group LVMH. And if they can't tell either way, then why bother worrying about it in the first place?

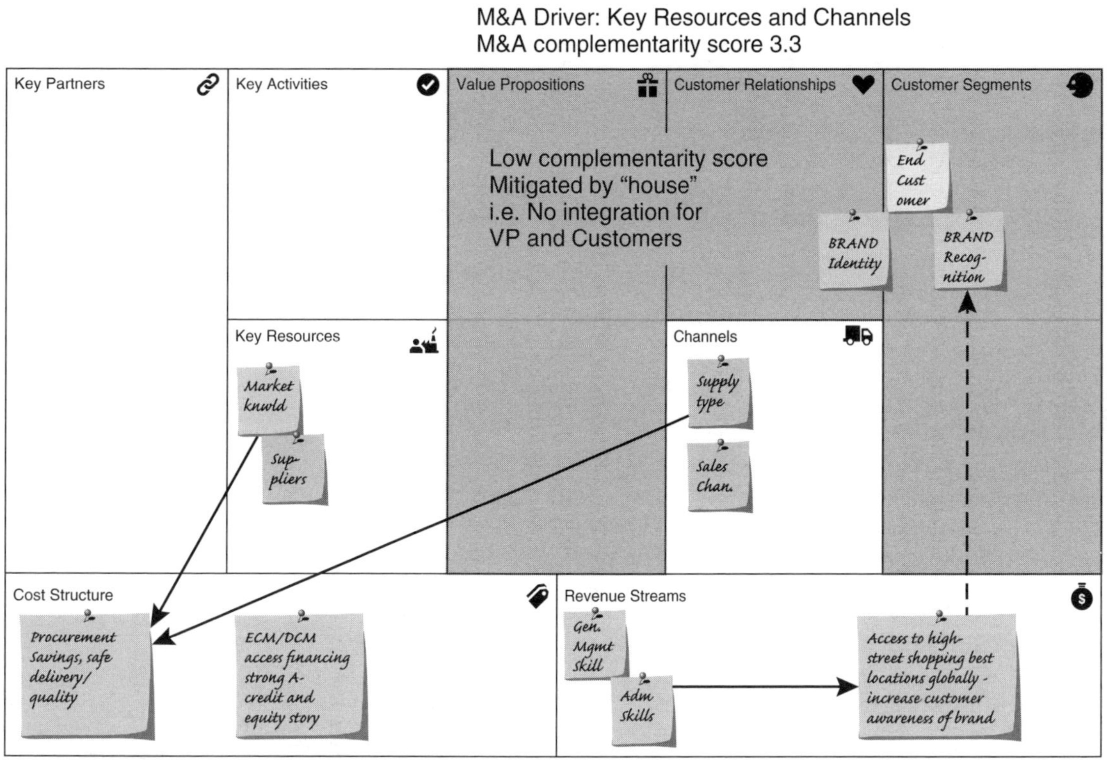

Figure 6.3 LVMH business model canvas

Table 6.6 RI3: Key activities and resources—LVMH

Research Insight title	Maintaining cultural heritage whilst being driven by key resources and activities
Case Insight to discuss	Louis Vuitton
Context from Peter Secher	Key resources and channels fully leveraged whilst insisting on keeping same customer segments, allowing acquired firm to continue building its own value proposition and keep its customer relations fully intact as 'House of LVMH'—M&A complementarity score 3.3
Research Insight content	**Point:** Target operational autonomy allows acquirers to rely on the strength of the target with minimal disruption. **Counterpoint:** Integration to eliminate redundancies and to achieve coordination in the organization is almost always needed, and must be managed with great care in highly autonomous targets if the business model should be preserved. **Contingency:** Integration and autonomy are not opposite ends of a scale, but can rather occur simultaneously in a single acquisition.
References	**Point:** Puranam, P., Singh, H., & Chaudhuri, S. (2009) Integrating acquired capabilities: When structural integration is (un)necessary. *Organization Science*, 20(2): 313–328.

(*continued*)

Table 6.6 *(continued)*

Counterpoint: Cording, M., Christmann, P., & King, D.R. (2008) Reducing causal ambiguity in acquisition integration: Intermediate goals as mediators of integration decisions and acquisition performance. *Academy of Management Journal*, 51(4): 744–767.
Contingency: Zaheer, A., Castaner, X., & Souder, D. (2013) Synergy sources, target autonomy, and integration in acquisitions. *Journal of Management*, 39(3): 604–632.

LVMH is the master of maintaining heritage and client journey with their brands. The company empowers the value proposition of any brand acquired and in the mind of the customers, the brand only gets stronger.

Too bad LVMH never got a majority of shares in Hermes because, from a one-eyed capitalist's point of view, it would be value-adding growth and just another slam-dunk business model-driven M&A deal. In M&A, it is important to never let feelings get in the way of a deal, but to focus on the business model fit instead (see Table 6.6).

CI4: The Global Brewer: Driving M&A Growth One Beer at a Time

Without naming any names, let's take a look at the M&A drivers for this global brewer in greater depth. The company is doing great. In fact, compared with its peers, it is probably one of the best in class. But many companies post excellent results and follow a path of solid, organic growth. In the brewing business, as in many other businesses, this is not enough to make you an industry leader in the long term. You have to do more than that to become great.

In order to create long-term value and ongoing growth, you also need a successful M&A strategy. This particular company excels when it comes to their M&A processes—they repeatedly do earnings per share accretive M&A transactions with a stringent focus on business model-driven M&A. They ask all the important questions: Why are we doing M&A? What sort of deals do we strive for? How do we cope with the various deal types? And what have we learned from previous M&A transactions?

In the investor community, there is widespread confidence in this company's M&A activities, and that is no mean feat! The clichés about corporate M&A are well known—at least 50% of all M&A deals are failures; and some 70–80% do not come close to the inaugural targets set out pre-deal. But this firm is unusual in that it has ignored these damning statistics and forged its own path. Not only does it successfully pursue business model-driven M&A, it has a proven track record of doing M&A mega deals.

Here is a company, at least to date, that seems to have the backing of its equity investors when doing M&A deals. In fact, some of its bond owners even see value in buying their corporate bonds immediately after M&A announcements, as the capital market is prone to over-react negatively in the wake of an M&A announcement. It is conventional wisdom among tier-1 investors to become fearful when an M&A transaction is announced, and who can blame them when the majority of these deals are failures?

However, this industry behemoth does not follow the usual rules of the game. Bet against them at your peril—this is one of the very rare companies which has a history of creating value in M&A deals time and time again.

At a big investor conference in Paris, all the talk was about industry consolidation—namely, who will buy next?

This brewing company was mentioned as an obvious next mover. Several equity investors were praising their M&A activity, and they openly speculated that the next M&A would probably *also* create value, just as the others have done. Not only would these investors stick to their equity holdings in this global consolidator, they were even considering an increase as they expected more value to be created.

So how do they do it? And what can other big firms learn from them?

The company's M&A activities can actually be summed up in just a few sentences.

Firstly, M&A deals are divided into four operating business models.

Secondly, each model comes with meticulous planning, making it possible to compare with historic deals in order to work out whether or not the deal is a realistic mission. The key people doing the deals are seasoned professionals—the Warren Buffetts of the brewing world! Like Mr. Buffett they very much like strong brands, but they also do deals with other interesting building blocks within their M&A business model. All deals have to offer 'intrinsic value' for the firm, as measured by return on net assets (RONA), a focus on total income generated from fixed assets and net working capital. This key focus is driven by the fact that brewing is inherently capital intensive, and many capital-intensive firms use these measurements to help them decide whether they should buy a company or not. The company does not have a fixed hurdle rate or a definition of excess return to weighted average cost of capital (WACC)/economic value added. The basic methodology is to ask: "Would we invest in this company's fixed assets right now?" and let everything else follow on from there. Of course, route to market and brand value also have a role to play, but using RONA is a vital measurement for capital-intensive businesses and in any case, the best M&A projects tend to have an interesting combination of high RONA and low WACC, adding value to the firm.

Table 6.7 CI4: A global brewer with business model drivers depending on the M&A deal archetype

10-year return on the global brewer's stock	
Initial investment 01/01/2007	€100
Would have grown to	€238.50
Yearly return	9.1%
10-year benchmark return Eurostoxx	
Initial investment 01/01/2007	€100
Would have grown to	€121.09
Yearly return	1.9%

Thirdly, everyone involved will know what needs to be done and why. The whole organization is punching at its full weight or above. Business development plays a key role in M&A for this brewer. Plant, property, and equipment needs are carefully measured out and challenged left, right, and center in any specific M&A transaction. Organizationally, this is a totally integrated process, with key people signing off on the business model and its individual building blocks.

Fourth and lastly, communication around the M&A deal is clear. Full company transparency makes it obvious for all stakeholders what is going on and why. Furthermore, how does this M&A deal compare to the ones we already did? How much can we pay for this company and when do we back out? A tried and tested M&A Playbook allows every single employee to monitor the deal at every stage and work out whether the M&A project is a realistic mission, and when to draw a line in the sand (see Table 6.7).

M&A Deal Type 1: Entering a New Market

The M&A drivers for this kind of deal are:

- *Good market position*. #1 or #2; in some cases #3 if there is reason to believe that it will grow to a higher position in the market.

- *Sizeable market share.* Preferably 25–40%.
- *Purpose.* Develop a premium segment.
- *End game.* Scale economies in order to achieve low-cost production per unit sold.
- *Business model driver illustration.*

M&A Deal Type 2: Existing Market Position

The business model driver for this kind of deal is clearly 'cost synergies,' and this is calculated in two ways (see Figure 6.4 as an example of M&A Deal Type 2):

1. *Cost.* A brewery is a highly capital-intensive business, and traditional wisdom has it that a capacity utilization of at least 75–80% is needed in order to break even on production (all other things being equal and in line with ordinary business, of course). In some of these deals in existing markets, it is possible to take out capacity from an existing brewer and produce your own excess capacity.

2. *Channels.* Or route to market, as it is preferably called by many fast-moving consumer goods (FMCG) companies. This business model driver is literally driven by a 'more on the truck if there is still some space' mentality. It is often used by pharma companies with narrow product lines when selling to hospitals or care centers, for instance.

    Business model-driven M&A that emphasizes cost synergies by taking out capacity or getting more into existing channels is probably one of the most predictable synergies you have in the world of M&A. And it is often achievable within a year if production allows it.

    The typical barrier for a leading global brewer, or other industry behemoth with leading market share, is normally 'just' the antitrust authorities—and these regulatory issues totally define the game. It is definitely a ceiling to accelerated growth on existing market positions. However, right up to this hard stop, most M&A deals come with a very high predictability indeed.

M&A Deal Type 3: Developing a New Company

Unlike the 'Type 2' deal, this type of M&A transaction comes with much more uncertainty. Still, for a global brewer with cash to boost capex, and many other brands to put into an existing—and quite effective—sales channel, it makes sense to pursue such M&A deals, even though they normally require a price premium. This must be paid by expected revenue synergies and, to some extent, more operational efficiency.

One such example comes from the Philippines, where this global company managed to acquire a local brewer. The local business did not have the market position required of a 'Type 1' deal, but the objectives were more or less the same. The company does not always find the production facilities or brand awareness in the right shape, but these things can be taken care of post-deal. Capex is increased, a more cost-efficient production set up, some premium brands from back home are introduced, together with the local brands, all with the purpose of developing the new brewing company. These deals do come in an M&A wrapping, but they probably share more DNA with 'business development' in the long run. They require years of experience in the industry, knowledgeable people who can lead the transformation, and, as we will also see with the next type of deal, the company must rely on a higher-than-usual staff retention to get these changes through.

M&A Deal Type 4: Greenfield

As the name may suggest, very few (if any) business model drivers are actually in play from day one on these deals. You have to start building factories and infrastructure, as you will typically not be present in the market from the start. The reason such deals are still M&A-related is often the fact that the company will buy some activities from a government. One such example can be found in a greenfield deal carried out on the Ivory Coast. Some bankers were hired to arrange an auction on state-owned breweries, and you are suddenly looking at a great greenfield opportunity. In these projects you really rely on locals, who can start the project but also have the right risk appetite as risk/reward is not as it is in the typical western economy. The brewer followed the advice from its own local management and introduced new brands based on penetration price politics. The low introduction price made the beer as successful as projected, and within a few months production capacity had been doubled. One of the key points to consider was the local consumer base. This country had a high population density, with more purchasing power, and the value proposition gave the people an opportunity to drink their own local produce.

In a surprising turn of events, the brewer then discovered that it could sell even more of this product in France, where there had been tremendous growth. The Ivory Coast has a strong French connection, and this gave the company further sales. While discussing this phenomenon an interesting point came up from the brewer—they said that they do not price in revenue synergies, although of course they are happy when this occurs. However, it is not built into the bidding strategy.

We have much more input from this brewer later in the book, where we take a deep dive into new M&A practices. However, it is perhaps interesting to mention that the normal horizon for transforming the business model from 'as is' to 'should be' is around 3–4 years. On greenfield operations, it is often 7–8 years, and obviously the company has to be more patient in its M&A approach, sometimes even playing with revenue synergies that may never materialize (see Table 6.8).

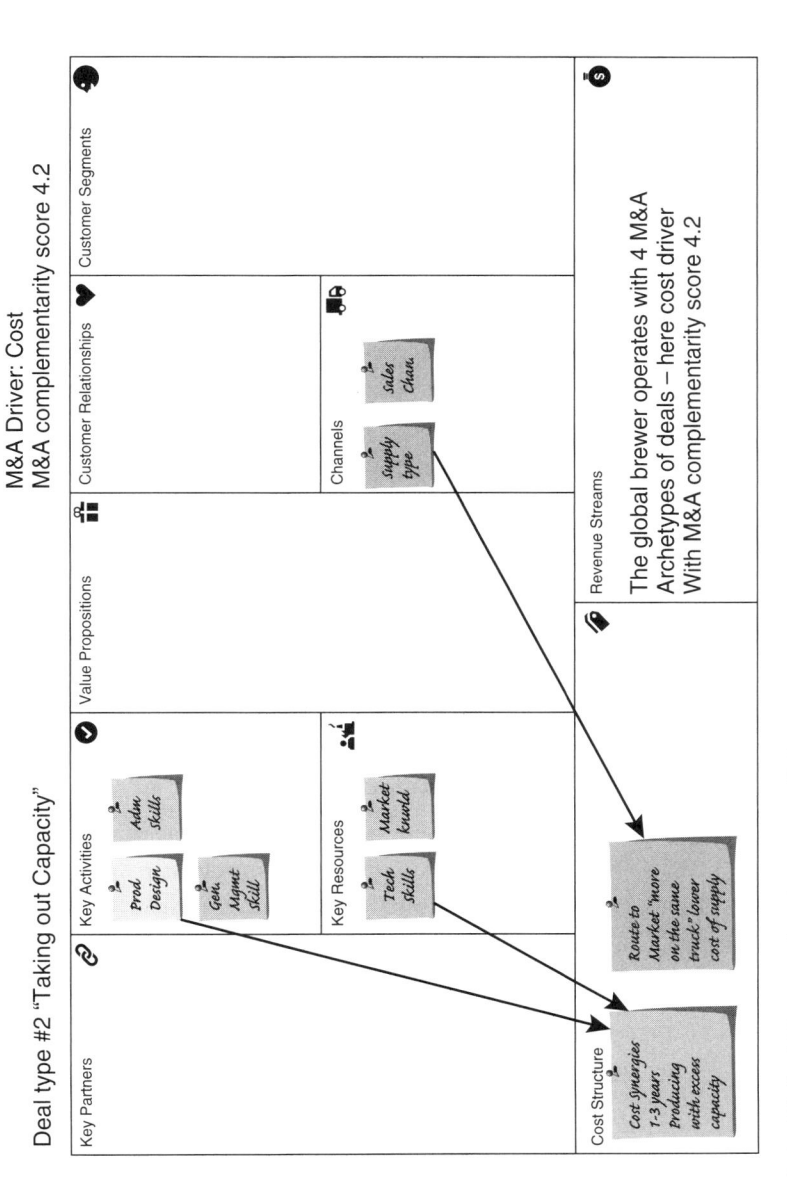

Figure 6.4 Global Brewer business model canvas

Table 6.8   RI4: Value proposition, customers, and cost

Research Insight title	Value proposition
Case Insight to discuss	The global brewer
Context from Peter Secher	Four M&A archetypes—in this example, cost as business model driver in M&A with a 4.2 M&A complementarity score—resource building block normally always replaced.
Research Insight content	**Point:**
	There are three different layers of capability needed: the capability to develop capabilities through acquisitions; the capability to conduct acquisitions; and to develop managerial capabilities to manage acquisition programs. Further, M&A needs clear guiding principles for how they should create value.
	Counterpoint:
	Based on previous experiences, firms might become superstitious or transfer and incorrectly apply prior experience without sufficient adaption. Further, vagueness or mixing principles might destroy value.
	Contingency:
	Positive or negative effects of acquisition experience depend on in-domain and cross-domain experience transfers, and sufficient experience to know when prior experience can be applied appropriately.
References	**Point:**
	Laamanen, T., & Keil, T. (2008) Performance of serial acquirers: Toward an acquisition program perspective. *Strategic Management Journal*, 29: 663–672.

Table 6.8

Bower, J.L. (2001) Not all M&As are alike – and that matters. *Harvard Business Review*, March: 93–101.

Counterpoint:

Zollo, M. (2009) Superstitious learning with rare strategic decisions: Theory and evidence from corporate acquisitions. *Organization Science*, 20(5): 894–908.

Contingency:

Bauer, F., Strobl, A., Dao, M.A., Matzler, K., & Rudolf, N. (in press) Examining links between pre and post M&A value creation mechanisms – Exploitation, exploration and ambidexterity in central European SMEs. *Long Range Planning*.

Haleblian, J., & Finkelstein, S. (1999) The influence of organizational acquisition experience on acquisition performance: A behavioral learning perspective. *Administrative Science Quarterly*, 44: 29–56.

CI5: Danaher: The Importance of Teamwork

We have just been through a few examples of business model-driven M&A where 'cost' is the M&A driver. Most people can get on board with the idea of business model-driven M&A when you can document the value creation within 1 or 2 years. For instance, it's safe to assume that customers will be using the same locks/entrance systems and drinking the same beer, at least in the short term. Meanwhile, the companies selling those products are able to use M&A to reduce the cost per unit. A cost-centric business model driver is also very easy to explain to staff, thereby making it much easier to create a sense of belonging in the M&A project for people in both organizations.

But 'cost' isn't always the most important business model driver. 'Key activities' can be a very successful M&A driver for many brand names. So far, we have shown examples of a few successful corporate acquirers who understand the importance of business model-driven M&A, strongly supported by good leadership and internal processes. This next example is a global company where business model-driven M&A is deeply rooted in the firm's key activity.

Danaher is unusual among its peers because its entire business model-driven M&A approach is fixated on this one driver—key activity (see Figure 6.5). And for Danaher, that key activity is manufacturing. This is a company which makes products across a number of different industries including life sciences, industrial techniques, and medical diagnostics. The company produces everything from dental equipment to marine exploration instruments. However, despite the variety of its products, the overall goal remains the same—to continue to manufacture quality products.

The Danaher Business System (DBS) approach is almost like a refined version of the Kaizen methodology—the Japanese discipline that strives for constant improvement. In our view, a key part of the DBS methodology is to act more like a conglomerate than a single business. This form of ownership normally makes equity investors rather reserved, as they prefer to diversify their investments themselves, as opposed to letting a conglomerate spread itself over many kinds of businesses and risk over-diversification. With just a few exceptions (e.g. General Electric), the world has not seen many conglomerates delivering a high total return to their shareholders, and there are even fewer examples of conglomerates acting as serial corporate acquirers across several industries (see Table 6.9).

The French entomologist August Magnan once claimed that the flight of the bumblebee was scientifically impossible, as the insect does not have the capability in terms of wing area or flapping speed. Yet, bumblebees still fly. Danaher is the bumblebee of the M&A world—it

Table 6.9 CI5: Danaher with key activity (DBS) as business model driver in M&A*

10-year return on Danaher stock	
Initial investment 01/01/2007	$100
Would have grown to	$292.20
Yearly return	11.3%
10-year benchmark return S&P 500	
Initial investment 01/01/2007	€100
Would have grown to	€195.40
Yearly return	6.9%

*Information based on Harvard Business School Paper No. 9-708-445 (11/30/2015) and various equity reports. No interviews or comments conducted with Danaher itself.

shouldn't work, yet it does. Conglomerates aren't supposed to offer value for shareholders, yet Danaher delivered above-average returns during our research period.

Even within our M&A Elite, Danaher's success doesn't make sense. Compared with the other M&A Elite companies, Danaher has a very low M&A complementarity, and the chances of M&A success normally increase with higher complementarity between the business model of the acquiring firm and that of the target firm.

Yet, despite all expectations to the contrary, Danaher 'flies' when it comes to M&A. This is largely down to the fact that Danaher considers its home-grown Kaizen-esque DBS to be a key activity in the firm, and the sole business model driver in its M&A Launchpad.

The only complementarity between the portfolio firms is the fact that they are all manufacturing firms, but with hugely different product-market building blocks. The value chain also varies significantly between each portfolio firm in areas such as after-sales service, suppliers, and sales channels.

You may have heard about some of the companies owned by Danaher:

- Fluke Networks
- Gilbarco
- Veeder-Root
- Hach
- Lange Trojan
- KaVo
- Gendex
- Radiometer
- Leica Microsystems
- Pelton & Crane
- Kollmorgen
- Portescap
- Dover
- Videojet
- Accusort Craftsman
- Matco

In practice, there are very few complementarities between these firms, but some common traits include a global market share roughly between 10% and 30% and a market position of #1 or #2 in their various fields.

So, in order to figure out how the company 'flies,' we took a closer look at the DBS system, which revealed a highly supportive leadership team, only going to prove how important people are in the M&A Formula.

No Red Tape in Danaher, but Magic Tape

The Radiometer acquisition is just one example of the Danaher method in action.

Following Danaher's acquisition of Danish blood-sampling firm Radiometer, one priority was to identify operational efficiencies by splitting the processes into smaller parts and strictly identifying all value streams. Almost immediately, a lot of low-hanging fruit was identified. One of the most obvious areas for improvement was a specific plastic part that took a little longer than 20 minutes to produce, but had an internal travel time of 18 days.

What does that do for the shareholders then? A lot. It increases cash conversion immediately. Higher free cash flow means higher valuation. Stock-listed firms like Danaher will immediately create value for their shareholders when they increase cash conversion. Needless to say, if a company reduces its time on inventory, then working capital is lower, and that is a unique recipe for M&A success. On top of this, if you are able to increase gross margins then you have very strong synergies in your business model-driven M&A.

This lean system was adopted by Danaher long before it got cool—1987. Ever since, the company has worked hard to make production more efficient and it always starts with people.

A 2010 article in the *Harvard Business Review* (HBR) found that Danaher was an expert in efficiency. For instance, in a typical Danaher factory, people will use tape to indicate where machinery, raw materials, and even trash cans ought to be placed for maximum efficiency.

This is just one facet of the DBS approach. Before proceeding with any acquisition, Danaher must be convinced that the target company will be able to abide by its tried-and-tested rules. If a company is not willing to use the 'magic tape,' it will not be a good business model fit.

Danaher is not your average M&A Elite player. It could not be described as a strategic buyer but, on the other hand, it is not a pure financial sponsor either—like a private equity firm. Instead, it is a vast company with more than 63,000 employees, which promotes transparency from the very top to the very bottom. It is the perfect example of the efficacy of having an M&A Deal Committee.

This approach has seen the company deliver a compound annual growth rate (CAGR) of over 25% by pursuing business model-driven M&A in line with our M&A Formula. According to the HBR study, the firm tends to look for low-profile industrial firms, with a market size of $1bn or more, and 5–7% growth. Most importantly, Danaher must be able to apply the DBS approach to the target firm. This is the company's version of our Goldman Gates, and it is the secret behind its unlikely success.

Once you understand their formula, you realize that the firm hasn't got to where it is today because of luck. Its power lies in identifying optimization opportunities and then executing them. This is its key activity. It is not down to 'magic powder' from the M&A department or a stroke of genius from an external advisor.

The consistency Danaher delivers in its business model-driven M&A is something any company can and should do—but they don't. According to a McKinsey article, most companies allocate the same capital to their business model activities every year.

It is exactly this corporate inertia that has inspired Danaher to be different. Any company acquired by Danaher has, in the eyes of Danaher, a totally illogical approach to capital and a 'we always did it this way' mentality.

Danaher's approach to every new portfolio company will always be non-intrusive, non-judgmental, and unemotional, so that this sort of corporate cognitive bias does not influence the deal. Instead, Danaher is purely focused on the opportunities for optimization, which is a vital part of the company's key activity of manufacturing (see Table 6.10).

M&A Driver: Key Activity (homegrown DBS concept)
M&A complementarity score 1.5

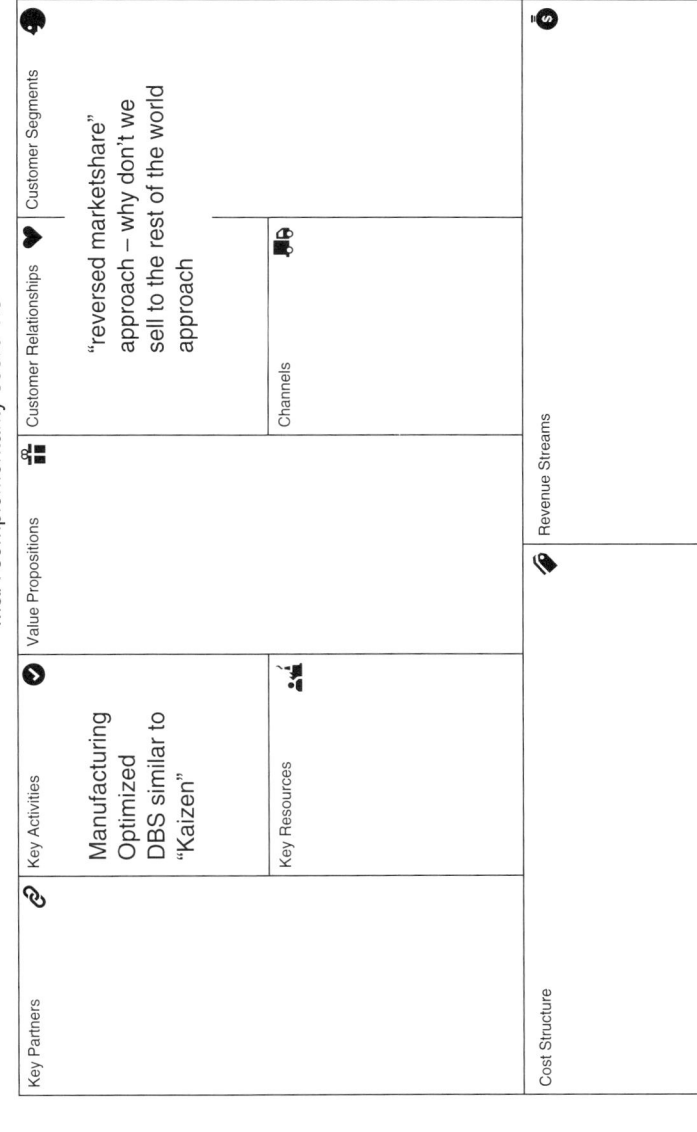

Key Partners	Key Activities	Value Propositions	Customer Relationships	Customer Segments
	Manufacturing Optimized DBS similar to "Kaizen"		"reversed marketshare" approach – why don't we sell to the rest of the world approach	
	Key Resources		Channels	

Cost Structure	Revenue Streams

Figure 6.5 Danaher business model canvas

Table 6.10 RI5: Danaher with key activity as business model driver in M&A

Research Insight title	Danaher with key activity as business model driver in M&A
Case Insight to discuss	Danaher
Context from Peter Secher	The home-grown management methodology DBS as key activity with an M&A complementarity score of 1.5. Despite the relatively low M&A complementarity, the company still repeats its success in M&A.
Research Insight content	**Point:** To find sources of competitive advantage, firms need to look beyond product markets to examine firm-specific activities and synergy that can be found in value chains. Further, value derives from the active transfer and sharing of resources and capabilities. **Counterpoint:** Changing existing processes and structures disrupts employees and thus might lead to productivity losses. Further, vertical integration in the value chain can create additional competition with suppliers and customers.
References	**Point:** Bauer, F., King, D.R., & Matzler, K. (2016) Speed of acquisition integration: Separating the role of human and task integration. *Scandinavian Management Journal*, 32: 150–165. **Contingency:** The transfer and sharing of resources requires change that must be managed, and coordination mechanisms are not free of constraints. For example, the stage of the industry lifecycle is an important antecedent for the applicability of integration measures.

Table 6.10

Counterpoint:

Paruchuri, S., Nerkar, A., & Hambrick, D.C. (2006) Acquisition integration and productivity losses in the technical core: Disruption of inventors in acquired firms. *Organization Science*, 17(5): 545–562.

Contingency:

Bauer, F., Dao, M.A., Matzler, K., & Tarba, S.Y. (2017) How industry lifecycle sets boundary conditions for M&A integration. *Long Range Planning*, 50(4): 501–517.

CI6: FrieslandCampina: A Merger of Equals or an Impossible Utopia?

We would all like to believe that each merger will bring two organizations together to create one unit of happy, productive employees. A merger of equals.

Every time a new merger is announced, we hear the same hopeful phrase being repeated again and again—"this will be a merger of equals." Yet this is rarely ever the case. I have been involved in quite a few mergers over the years, most of which could be described as successful. However, since the majority of these mergers occurred within the banking industry, this success was usually defined in terms of value creation. While the idea of creating a 'merger of equals' was nice on paper, it proved to be almost impossible in reality. One company will always come out stronger than the others, and each entity will be forced to make some sort of compromise along the way.

The first really big merger I was involved in was the merger between Handelsbanken, Provinsbanken, and Danske Bank in April

1990—a merger that created the largest Scandinavian bank. As ever, the deal was announced as an equal merger, despite the fact that both the CEO and the chairman were appointed from Danske Bank. This is a common occurrence in these so-called 'mergers of equals,' where one company is usually able to retain more senior management than the others. In fact, one could argue that this is the healthiest course of action. Is it not better to have one strong leader who can get things done? Yes, but when one company takes the lead post-merger, it can cause divisions and issues further down the line, even years later.

Back in 1990, 'Danske Bank' became the new name for the merged banks, and the Handelsbanken logo was used for a short time thereafter. As the smallest brother in the family, Provinsbanken totally vanished.

While people soon got used to the new brand—the new 'Danske Bank' brand—something interesting started to happen among the employees. Even several years after the huge merger, they would refer to the bank's operations as being red (the color of the Danske Bank logo), blue (the color of the Handelsbanken logo), or green (the color of the Provinsbanken logo). They could let go of the names, but not the history. On occasion, people would be heard bad-mouthing the new bank for not approaching certain decisions in a 'blue' or 'green' way. There was criticism of 'green' leaders, 'blue' systems, and 'red' processes. Every time a major decision was being made, there would be a ridiculous discussion over colors rather than simply coming together with the sole goal of making the bank stronger in the market. This continued to be an issue for at least 5 years.

It was then that I realized that a 'merger of equals' is a utopian concept within M&A. It could never happen.

But a few years later, I was proved wrong.

FrieslandCampina

Numbers often speak for themselves. Prior to the merger, the owners of Friesland Foods and Campina were lagging the milk price of their biggest nearby competitor by ca. €1.50 in the year 2008—this year the guaranteed milk price was €37.22. They paid farmers 4% less for the milk they delivered every day, compared with a nearby country in Europe. However, only 2 years after the merger, cost as the business model driver started to pay off, as both companies reaped the identified cost synergies (in the third year the newly merged FrieslandCampina paid for the first time a higher price than their competitor, namely €2.32). Just to put things in perspective, the milk price paid to farmers was now suddenly 7% higher—coming from 4% lower in three years! This can be compared with EPS on stock-listed companies and is quite an improvement in only 3 years. Many stock-listed firms could only dream about such an improvement. Still, the merger continued to pay off. 'From push to pull'; started to work, with increased branding meaning improved value proposition, stronger client relations, and a broadening of the global client base. FrieslandCampina developed itself into a strong company with an emerging market exposure of about 33%. Exposure here is a positive, and Western Europe continued to offer low-growth opportunities for dairy firms, with the single most negative driver being a growing population of the elderly. You may remember drinking more milk as a child than what you consume in your coffee

(continued)

(continued)

or tea today ... Add to that the strong demand in emerging markets, where most of the world's children are born today. They demand an incredible amount of baby food, which is one of the highest gross margins available in the dairy business as of today.

If we run the same traditional investment analysis as investing in stock-listed shares, which is of course not possible as milk itself is a perishable product as opposed to a financial investment, still the farmers of FrieslandCampina were paid significantly more than their nearest competitors—a gap that increased to almost €3.00. The year of the merger, the guaranteed price paid to farmers was €37.22 (here the base investment). In the coming 10 years they were paid an accumulated ca. €23.00 above other dairy farmers, who 'didn't hold shares' in FrieslandCampina—therefore ending up with a future value of the higher cash flow of ca. €60.00. This is only about 5% per year but compared to other dairy firms well above zero or even worse, as most dairy firms have actually been struggling in the past 10 years due to the abolition of the European Union's milk quota system in 2015, which means increased production of dairy products way above demand globally.

In 2011, I joined FrieslandCampina. Three years earlier the company had been formed through a merger between Friesland Foods and Campina, and by 2011 it had more than 22,000 employees, a turnover of almost €12bn, and products represented in some 120+ countries around the globe. I became head of the M&A department and one of the first things I was told was that this huge company was—guess what—a 'merger of equals.' Not again, I thought ...

So I started to wait for the inevitable conflict, one employee bad-mouthing another because he or she was acting like another 'color,' or a department complaining about a stupid IT system, which wasn't as good as the previous setup. But it never happened.

I didn't give up. By now I was certain that a 'merger of equals' could never exist, and the sooner I proved my theory, the sooner I could really get to work. I started to rigorously investigate the dynamics of this new company. Who had taken control? Where did the CEO come from? Which company gained and which company lost?

In fact, following the merger, both companies had asked their respective CEOs to step down in order to give the merger a greater chance of success. They appointed a new CEO—Mr. Cees 't Hart—who was not connected to either of the previous two entities. As a result, the merger had succeeded in creating a fresh new company which combined the strengths of each brand without causing friction between the employees. I was deeply impressed.

I remember identifying three key reasons behind this 'merger of equals,' which I vowed to carry with me for the rest of my career.

Firstly, and perhaps the single biggest reason behind the success, was the brave decision by the respective chairmen to agree on a completely new CEO with no legacy in the merged companies. You could simply not associate a color or culture with this new CEO.

Secondly, Mr. Cees 't Hart, in my view, was extremely good at bringing people together to ensure that everyone was working towards the same goals. This is no easy task.

Thirdly, this was a textbook example of business model-driven M&A (see Figure 6.6), with a clearly identified M&A driver in the immediate aftermath of the merger: reaping cost synergies. No longer would two trucks with two different logos drive side by side between adjacent fields with milk or other produce. Changes like this created almost €200m in cost savings, or almost 5% of the former Campina's

Table 6.11 CI6: P + VP

10-year milk price* (investment comparison)	
Initial investment 01/01/2007	€37.22
Would have grown to	€60.50
Yearly return	5.0%
10-year benchmark return competing dairy firm	
Initial investment 01/01/2007	€37.22
Would have grown to	€40.30
Yearly return	0.7%

*As an unlisted company, there is no publicly available data on FrieslandCampina's share price. However, the company's success is evident from the evolution of its milk prices. Since the merger of Friesland Foods and Campina, the firm's farmers are being paid more for their milk, so volumes have gone up and profits have risen.

overall turnover. This is a huge achievement for any company, and quantifies the success of the merger in terms of both value creation and brand equality.

I often use this merger as a watermark for other cost-synergy deals, where the M&A driver was cost cutting. The newly merged company focused on this driver day in and day out, and the results speak for themselves (see Table 6.11).

The key business model driver for the merger between Friesland Foods and Campina was cost cutting. However, it wasn't long before another business model driver was identified. By hiring a brand-new CEO, the firm was able to strengthen its brand via what the newly merged company would call 'from push to pull.' It was clear to everyone involved what the business model drivers were in this newly merged entity. If you asked anyone in the company to name the key drivers of the merger, they would say 'cost cutting' and 'stronger branding.'

What Was Behind the Timing of the Merger Talks?

According to Mr. Hart, a series of external circumstances added some urgency to the M&A talks, encouraging both parties to begin inaugural discussions sooner rather than later.

Firstly, one of the companies was struggling financially, and was veering close to bankruptcy.

Secondly, Friesland Foods had just witnessed the failed merger between Campina—its biggest rival—and Arla Foods, another major competitor. In the immediate aftermath of this failed deal, Friesland Foods spotted a window of opportunity and immediately went for it.

The talks soon turned towards M&A driver 'scale economies,' and it became clear to both parties that they could be looking at large cost synergies. McKinsey was hired to identify all possible savings, and concluded that "the total cost synergies could be larger than €100m yearly." To put this in perspective, the smaller of the two firms (Campina) had at that point a yearly turnover of about €4bn, so the cost savings could be higher than 2.5%.

This is not a bad level by any means, many companies would love to reach such high cost synergies in their M&A deals! Still, what was actually delivered in total cost synergies over the next 5 years was almost double the inaugural target. This came as a (pleasant) surprise not only to McKinsey, but also to the company itself, and represented a high watermark in terms of successful business model-driven M&A.

To secure these potential savings in the newly formed company, both chairmen agreed to look outside the companies for an 'unknown' CEO who could effectively reap the cost synergies and create a new business model. They chose Cees 't Hart. However, they wanted to be very careful not to alienate any of their existing senior managers, who brought with them a wealth of experience from both companies.

So, before the merger won anti-trust approval from Brussels, a long list of 70 names was drawn up, featuring top executives from both Friesland Foods and Campina, as well as a few senior newcomers. This 'top 70' became an informal advisory group, who met regularly to discuss company strategy and to ensure that the merger was running smoothly for everyone involved.

Going Far Together

Following his appointment as CEO of the newly merged Friesland-Campina, Mr. Hart was reminded about an old African proverb: "go fast, go alone; go far, go together." This would come to resonate with him as he began work on this merger of equals. To Mr. Hart, 'going together' meant creating a strong team. He told me how he was greatly inspired by a professional climber who later turned mountain climbing into a commercial discipline. He remembered the first words which he uttered upon arriving at base camp on Mount Everest: "we are a team in name only … over the coming eight weeks we will train together and learn to become one team so we can climb Mount Everest!"

With these words ringing in his ears, Mr. Hart gathered together his 'top 70' team on a European mountain called Schilthorn. There, he started sharing his idea of creating one team and a new culture within the merged company. He wanted everything to be completely aligned throughout the whole organization, and he was focused on a bottom-up roll-out focusing on the key M&A drivers which had been identified in the merger. A buy-in from 'top 70' meant creating brand defenders of the new corporate strategy with a strong focus on cost synergies and branding.

You may think by now that this approach was perhaps a little too friendly for a major, newly merged corporation, but from the very start of their discussions the caveat emptor was: "no silent disagreement allowed." The team stayed on top of the mountain until everybody had been given their chance to speak up, express disagreement, or (in most cases) buy into the new strategy.

Mr. Hart wanted to create a feeling of belonging to this new company. Furthermore, he wanted to give his new employees a clear direction as well as purpose. All this was implemented in a new corporate strategy named 'Route 2020.'

The initial M&A drivers for the deal were crucial when it came to explaining the 'Route 2020' vision, particularly the cost-cutting driver. But some of the new drivers proved more challenging than others. The change 'from push to pull' was crucial in terms of cost synergies and branding, but in order to work, it had to be understood. Mr. Hart explained it in terms of milk sales, a sector with which both companies were extremely familiar. Originally, the owners and farmers had the view that the milk should be sold (or 'pushed') into the market, but the new organization was going to focus more on branding, creating a 'pull' effect whereby customers would actively look for FrieslandCampina brands in the supermarkets and elsewhere. To really create a sense of teamwork, a new company logo was agreed with the 'top 70' executives and officially unveiled just six weeks later. This distinctive logo—rainbow colors around a white center—is still proudly displayed on FrieslandCampina's products around the world. How many executives can walk into a supermarket and say that they were responsible for co-creating the logo on that pint of milk? This certainly helped create a sense of belonging and cohesion among the 'top 70,' and the rainbow logo came to represent the 'milk' of the company itself.

What Happened Next…?

In the years that followed, not only did the Campina brand become a Dutch champion, but FrieslandCampina became the most sustainable company in Holland, winning prizes all over the world for its branding and efficiency. Since the merger, the new company has been able to offer an even higher payout to its owners and farmers, further cementing the fact that the merger was a success. The cost synergies were reaped at an even higher ratio than expected, but in my opinion it was the 'push–pull' conversion which was the company's greatest achievement. These revenue synergies are notoriously hard to achieve in corporate M&A, but somehow this team did it.

Of course, there were also some stumbling blocks along the way. After they came down from that mountain, a handful of executives either disagreed with the direction of the new company or simply changed their mind. While a few people chose to move on, in the end there was a strong buy-in of almost 95% across the 'top 70.'

Despite this, Mr. Hart maintained a strong focus on changing the company, and within 18 months almost 50% of the original 'top-70' had moved on to new roles. Experts were brought in for key areas such as HR, IT, Research, Procurement, and M&A, and the company stayed true to its mountaintop vision, growing stronger along the way.

The Key Learnings from this 'Merger of Equals'

Firstly, identifying strong business model drivers is essential in successful M&A. Not only does this create a clear target, but it can also lead to a transformation as it becomes so much easier to actually create transparency across the entire company. We call it 'punching above the organizational weight.' In this case, the business model drivers were cost synergies and stronger brands, but the new 'top 70' was immediately on board because they felt a sense of belonging and also had a clear direction and purpose. They even outperformed the inaugural expectations of the M&A talks.

Secondly, never forget that this is a 'people' business. The Friesland Foods and Campina chairmen focused on creating a new culture by

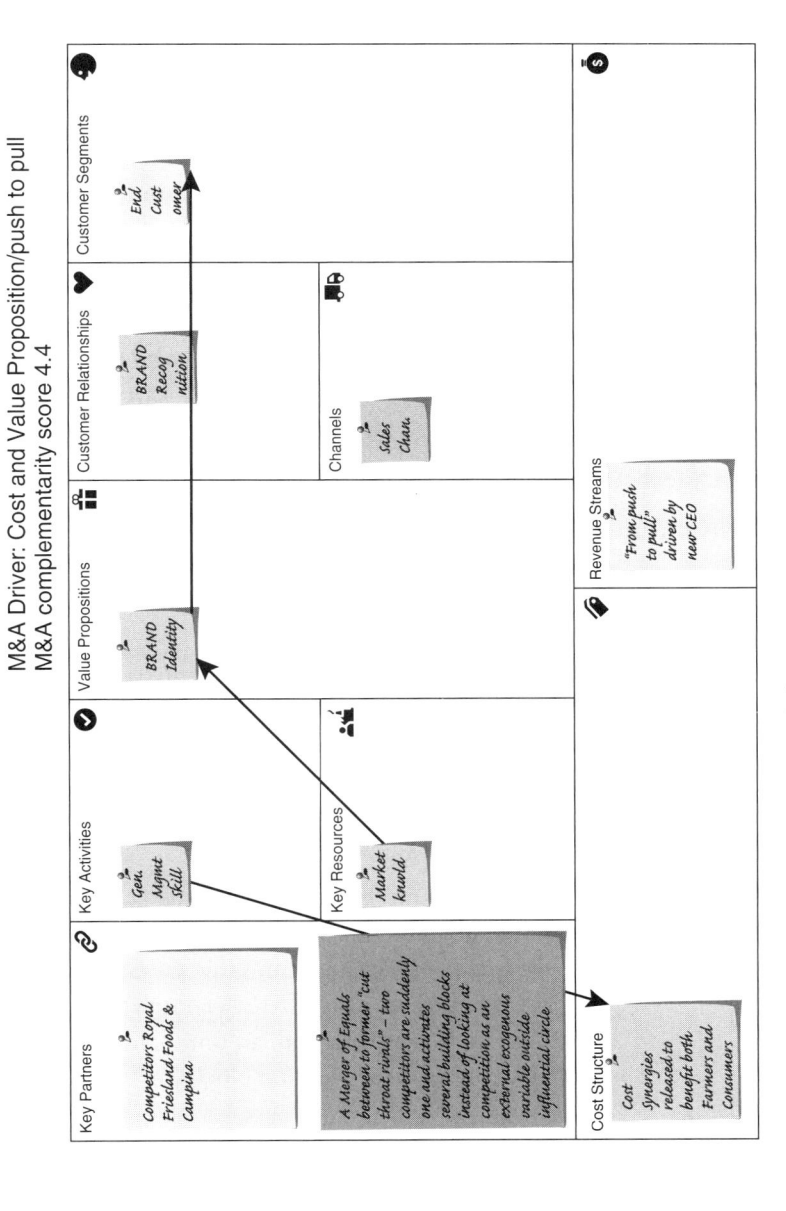

Figure 6.6 FrieslandCampina business model canvas

bringing in a brand new person. This was a brave move, but it also turned out to be a stroke of genius. They actually brought in a person without any experience in the dairy sector, but he did have experience when it came to dealing with people. This allowed him to focus on bringing people together within the new organization, and building a strong company culture.

Thirdly, strong leadership is essential. This may be an obvious point, but it is not always straightforward in M&A. Poor leadership is one of the main reasons why there is such a high failure rate in global M&A. It is up to the CEO and the chairman to deliver the message throughout the organization, so that everyone knows why they are doing M&A. Good leadership means identifying clear M&A drivers, delivering this message across the organization, and making sure that you are on a realistic mission from start to finish.

It turns out that a merger of equals can exist after all (see Table 6.12).

Table 6.12   RI6: Cost in short term and value proposition in long term

Research Insight title	Cost as short-term business model driver and value proposition as long-term value driver
Case Insight to discuss	FrieslandCampina
Context from Peter Secher	FrieslandCampina managed to reap cost synergies above anyone's expectations. Add to that a successful improvement in value proposition in a 'push–pull' conversion. M&A complementarity score on merger 4.4, with an unusually high score of 4.5 in resource building blocks.
Research Insight content	**Point:** Larger deals offer greater synergy potential as market power increases and costs can be reduced.

Table 6.12

	Counterpoint: Larger deals require greater managerial attention. The core internal risks are political behavior, in-group and out-of-group biases, and many more. Externally, larger acquisitions trigger competitor reactions and retaliation, and acquirers should be prepared for this. **Contingency:** Value does not simply occur. Mergers of equals must be managed properly (even if cultures fit and the deal makes sense for employees), and provide organizational clarity with communication and fast managerial action required.
References	**Point:** Kitching, J. (1997) Why do mergers miscarry? *Harvard Business Review*, 45: 84–101. **Counterpoint:** Dao, M.A., Strobl, A., Bauer, F., & Tarba, S.Y. (2017) Triggering innovation through mergers and acquisitions: The role of shared mental models. *Group and Organization Management*, 42(2): 195–236. King, D.R., & Schriber, S. (2016) Addressing competitive responses to acquisitions. *California Management Review*, 58(3): 109–124. **Contingency:** Dao, M.A., Bauer, F., Strobl, A., & Eulerich, M. (2016) The complementing and facilitating nature of common ground in acquisitions – why task and human integration are still necessary in the presence of common ground. *International Journal of Human Resource Management*, 27(20): 2505–2530.

CI7: ASSA ABLOY: The Highest Total Return to Shareholders in Our Research Period (1st Jan 2007–1st Jan 2017)

Despite conventional wisdom that at least 50% of all corporate M&A activity ends in failure, we have been able to show you six companies that consistently deliver value to their shareholders through vigorous M&A activity.

We would now like to introduce you to the seventh company … And this company also happens to be number one in our M&A Elite when it comes to business model-driven M&A (see Figure 6.7).

But before turning to ASSA ABLOY, let us first kill off a myth that is constantly plaguing the world of corporate M&A.

This so-called 50% or more failure rate appears everywhere. But if the odds of success are 50/50, why don't companies save themselves a lot of time and money and just flip a coin instead?

Because this myth is based on the law of averages, and that really sucks.

In every industry, you will have a very small group of elite leaders, and then everyone else. The elite players are not particularly interested in the performance of the average players, they just want to know what their peers are doing. Do you think Serena Williams studies footage of amateur tennis matches in order to improve her game? Of course not. She enlists the help of the world's leading coaches and tennis experts.

Likewise, I am not sharing techniques on how the average company avoids M&A failure, I'm showing you what the M&A Elite are doing.

There is another prevailing myth about corporate M&A—the idea that M&A activity doesn't create value for shareholders. Again, the M&A Elite are rejecting this theory.

Against all odds, these seven companies can somehow create value from M&A activity, again and again. In total return analyses of these companies we have been able to prove how much better they

Table 6.13 CI7: ASSA ABLOY—the winner of Global M&A Elite with M&A as key activity in the business model itself!

10-year return on ASSA ABLOY stock	
Initial investment 01/01/2007	Skr 100
Would have grown to	Skr 427.90
Yearly return	15.6%
10-year benchmark return Eurostoxx	
Initial investment 01/01/2007	€100
Would have grown to	€121.09
Yearly return	1.9%

are at returning money to their shareholders than, for instance, their domestic stock index.

Over a 10-year period, each one of these companies has greatly outperformed their domestic index by up to 10% every year. If you don't think that sounds like a lot, you don't know a lot about investments.

What separates the top 1000, top 100, or top 10 companies are small margins. Any company that delivers an annual return just 1–2% above their index over a 10-year period is a fantastic company. It is hardly surprising to learn, therefore, that the M&A Elite shares have performed way above any index and most individual companies.

If you had invested €100 1st January 2007 in our M&A Elite Shares, then 10 year after you would have had €236.11 (1st Jan 2017). A similar amount €100 invested in a market portfolio such as the Eurostoxx50 would only have grown to €121.09 in the same time period. Finally, an investment of Skr 100 in ASSA ABLOY would have left the investor with Skr 427.90 i.e. a far higher return. So, let's take a closer look at the winner of our M&A Elite index (Table 6.13).

Acquisitions in ASSA ABLOY

ASSA ABLOY is the world's largest lock manufacturer, with more than 45,000 employees and assets in excess of €8.95 billion. The group continuously makes acquisitions, and even sets goals for its M&A activity—they must contribute to 5% sales growth every year, and deliver an earnings before interest and taxes (EBIT) margin of 16–17%.

In fact, the company's business model has M&A listed as one of its key activities. The annual growth is divided into organic growth (i.e. running the business) and the other growth driver, M&A (i.e. acquiring other business models and adding them to the business model).

The company is itself the product of a 1994 merger, between the Finnish firm ABLOY and the Swedish firm ASSA. Since then, the company has demonstrated that acquisitions can become important building blocks in the successful growth of a company.

The group has seen unusually rapid growth throughout the years, with a CAGR of 16% between 1994 and 2015, and a sales increase from €300 million to €7.09 billion in just 21 years, while total sales have increased by 2,300% since 1994. Operating income, excluding items affecting comparability, has increased by over 7,100% from €16.3 million to €1.16 billion since 1994.

This success is not merely the result of successful positioning during a technology shift and the emergence of a new industry, or of revolutionary changes in demand or market conditions. Rather, it can be viewed as the result of an insightful analysis by a management team who dared to embrace change and who, right from the start, made a number of accurate and intuitive decisions that led to an upward spiral of learning and culture.

While acquisitions have helped ASSA ABLOY to shape the market, they have also shaped ASSA ABLOY as a company. Through the company's efforts to meet new customer needs and explore new opportunities presented by technology, new markets, and—above all—new people with new ideas, ASSA ABLOY has developed an extremely empowering culture and a great openness to the companies (and people) the acquisitions bring into the group. This is clearly shown in the fact that many of the senior managers came into the company through acquisitions.

All of this makes ASSA ABLOY a unique organization—created by an acquisition and driven by the never-ending quest to create value.

What follows is an exclusive insight into the risk-taking and decision-making that made it possible for this firm to successfully acquire and integrate more than 200 companies since 1994.

Acquisitions as Corporate Building Blocks

Over the years, the board of directors and group management of ASSA ABLOY have used acquisitions to secure continued growth for the group. There has been a clear goal of delivering profitable growth, and the fundamental principle has been to 'build and leverage' existing businesses and competencies. Through many well-selected acquisitions, ASSA ABLOY has built strategic platforms and bolted on smaller companies. Each time ASSA ABLOY acquires companies and/or new products within different territories, the plan is to find the market leader or product champion even if it requires patience. Any bolt-on acquisitions will be intended to strengthen market shares and deliver margin improvements via synergies.

Acquisitions have been used to establish sales in new countries and add new products and sales channels. However, at some point in time the market share increases so much that future acquisitions become limited. As a solution to this constraint, new and different business areas have been selected and acquired. For instance, in 2003 the group acquired the Besam automatic door business, which facilitated an organic and acquisitive expansion into 'entrance solutions.'

Besam, together with other brands, has grown to become the company's largest business division (Entrance Systems), driven mainly through acquisitions. During the period 2008–2016, the Entrance Systems division saw its revenue grow from €311 million to €2.07 billion, a CAGR of 27%.

The ASSA ABLOY Way

The group management has delegated the responsibility for acquisitive growth to the divisional management of its five business divisions—three lock divisions, one entrance system division, and Global Technologies (which includes HID and hospitality business units). In each division, there is a dedicated M&A team, who together with business and product people search and select potential acquisition opportunities. There is a continuous search for acquisition targets, and the group's acquisition strategy is focused on three things:

- increasing geographical market presence;
- complementing the product range, so they can offer their customers innovative and complete door solutions;
- adding new technologies in key areas.

A number of acquisitions have been made in each area, and there is a well-defined framework for what is expected post-closing. For all acquisitions, there are clear financial goals—reach 20% return on corporate equity (RoCE) within 3 to 5 years of the closing date, as well as delivering EBIT at an ASSA ABLOY level. These goals are a prerequisite for each acquisition, since they allow ASSA ABLOY to generate the cash flow needed to fund future acquisitions.

However, before a buying mandate is given for any acquisition, there are three fundamental requirements:

1. There is a strategic fit to the ASSA ABLOY business model.
2. The current management continues in the business.
3. There is an earn-out based on company profitability.

All three requirements must be fulfilled, otherwise experience shows that the risk of failure is much higher, and the group management will not approve it.

As mentioned earlier, ASSA ABLOY develops its business by building and leveraging the business and competencies. Therefore, each acquisition is part of a plan and fits into the business model either as a new strategic building block or a bolt-on of market share, products, or technology. The acquisition targets must fit into the ASSA ABLOY business model, and the archetypes of acquisitions must add 'value propositions,' 'customer relationships,' or 'channels.'

ASSA ABLOY acquires companies with healthy growth and profitability, so they can instantly join the business network. On top of a good financial performance, ASSA ABLOY always identifies synergies, and these most often take place within the 'cost structure,' 'customer segment,' and 'revenue stream' parts of the business model.

Frankly, ASSA ABLOY usually doesn't calculate synergies in the 'revenue streams' because they are riskier and therefore less likely to be implemented. Consequently, the sales synergies are normally not included in the valuation (see Tables 6.14–6.16).

The ASSA ABLOY Acquisition Process

Over 22 years and more than 200 acquisitions, ASSA ABLOY has developed an acquisition process that involves group management twice only—the first time when they give the mandate to sign the LOI, and the second time when they give the mandate to sign the final purchase contract.

The work towards signing the LOI is information heavy, which means that a thorough review of company financials, management, and strategy is done, and it is compared to ASSA ABLOY's current business. The preliminary integration plan is developed and the synergies are calculated. During the due diligence investigation, all areas are thoroughly investigated and verified against targets.

Further, there is a clear M&A policy in place, which describes the roles and responsibilities of group and division staff.

Table 6.14 Expectations for each type of deal (subject to size)
Customer relationships and channels

Increasing geographical market presence	Large, strategic acquisitions (Nassau in DK in 2016)	Smaller bolt-ons (L-Door in BE in 2015)	Smaller bolt-ons (Lighthouse in US in 2016)
Increasing market share	+20%	5–10%	+20%
Sourcing savings (purchase power)	2–3pp EBIT improvement after 12–24 months	2–3pp EBIT improvement after 12–24 months	No
Reduce complexity in product design by using components, software, and technology available in the group	Yes	Yes	No
Optimize the manufacturing footprint	Yes	No	No
Reduce number of sales offices and subsidiaries	Yes	No	No

Leverage the ASSA ABLOY commercial network	Only to a small extent	Yes	Yes
		• Add ASSA ABLOY products to L-Doors product offering and thereby increase sales volume of existing products	
		• Add L-Door products to ASSA ABLOY sales channels outside Belgium	
Convert sales of competitor products to ASSA ABLOY products	No	No	Yes Increasing the manufacturing volumes in own factories

Table 6.15 Value propositions

Complementing the product range	Large, strategic acquisitions (Crawford in SE in 2011)	Smaller bolt-ons (Nergeco in FR in 2015)
Increasing market share	+20%	5–10% (in EU)
Adding products to the portfolio	Potential to become one-stop-shop supplier	
Adding manufacturing sites	Potential to optimize manufacturing footprint for the combined company	
Sourcing savings (purchase power)	2–3pp EBIT improvement after 12–24 months	
Leverage the ASSA ABLOY commercial network	No	Yes • Add ASSA ABLOY products to Nergeco product offering and thereby increase sales volume of existing products • Add Nergeco products to ASSA ABLOY sales offices

Table 6.16 Value propositions—technology

Adding new technologies in key areas	Large, strategic acquisitions (CEDES sensors in CH in 2016)
Time to market for new products	Reduction
Sourcing savings (in-house manufactured instead of external sourcing)	5–10pp GM improvement after 12–24 months

Patience...

Part of their success is saying 'no' to many acquisition opportunities that do not match the firm's fundamental requirements and financial goals:

- increasing geographical market presence;
- complementing the product range;
- adding new technologies in key areas;
- 20% RoCE after 3–5 years.

Group management has shown that it has benefitted from the continuous search for acquisition targets and made the decision when the right candidate was for sale—small or large. Patience also means waiting for the large acquisition opportunity before entering a new product area or new geographic area.

An example of this is the ASSA ABLOY Entrance Systems business in the USA. Entrance Systems has a three-product offering within automatic door solutions. In 2011 it acquired Crawford, a large player in Europe, but due to the product design, it did not make sense to export from Europe to the USA. Therefore, three companies (Albany, 4Front, and Amarr) were acquired to build a manufacturing base and significant market share. Each of the three companies had a market share north of 15–20% and a potential to expand. With these three companies in the Entrance Systems division, the company could serve US customers with the same product portfolio as in Europe.

Look Me in the Eyes...

An important fundamental for success is confirmation of the seller's continuous commitment. Once the deal is done, it is of high importance that the management team of the acquired business remains in office and continues to deliver according to the business plans and synergies.

Here, ASSA ABLOY commits a lot of resources to meeting and interviewing the current management team. The aim is to be sure that the sellers are excited about joining the bigger group and will make use of the commercial, technical, and personal opportunities post-deal. This might actually be the single most important issue of the entire process, and ASSA ABLOY has in several cases walked away from deals where sellers were not committed to the ASSA ABLOY business model fit.

EBIT Improvements

The group has a lot of experience in improving the profitability of its acquired companies. Most of these improvements are quite obvious and can be executed shortly after the deal is completed. Often, it is a number of small improvements that add up, but of course major restructuring programs are also on the agenda.

The continuous optimization work being done in each division is from time to time accelerated by corporate manufacturing footprint programs (MFPs), where corporate funds are made available to each division to make larger manufacturing restructuring projects. For example, consolidation of the manufacturing of locks, motors, etc.

The EBIT improvements are:

- Sourcing saving—a 2–5% EBIT improvement
- Office consolidation
- Subsidiary consolidation
- Logistic and distribution network savings—fill the trucks, consolidate warehouses
- Manufacturing footprint—close manufacturing units
- Increase manufacturing in low-cost countries (LLCs)
- Reduce complexity in product design and utilize technology in larger units

Convert sales of third-party products to ASSA ABLOY.

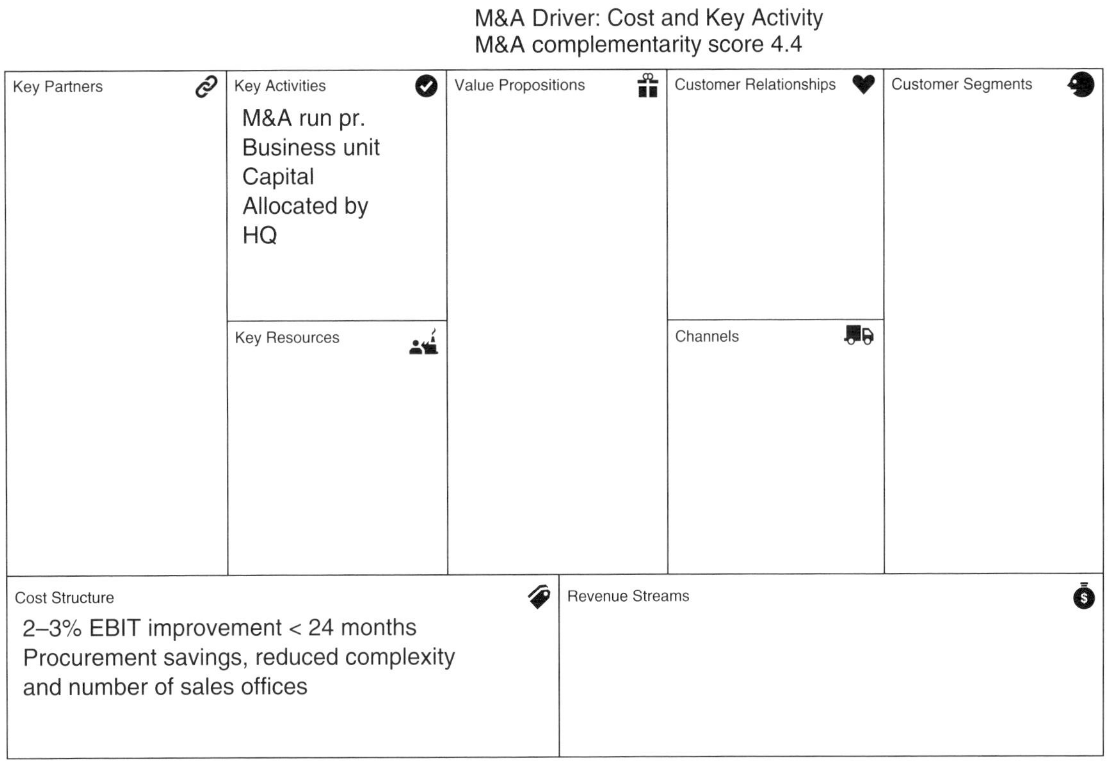

M&A Driver: Cost and Key Activity
M&A complementarity score 4.4

Key Partners	Key Activities	Value Propositions	Customer Relationships	Customer Segments
	M&A run pr. Business unit Capital Allocated by HQ			
	Key Resources		Channels	

Cost Structure
2–3% EBIT improvement < 24 months
Procurement savings, reduced complexity and number of sales offices

Revenue Streams

Figure 6.7 ASSA ABLOY business model canvas

How ASSA ABLOY Stacks Up Against the Business Model-Driven M&A Platform (Table 6.17)

Table 6.17   RI7: Business model-driven M&A at its best

Research Insight title	Business model-driven M&A at its best
Case Insight to discuss	ASSA ABLOY
Context from Peter Secher	M&A is a key activity, with a committed growth target of 50% of future growth, delegated to business units. M&A complementarity score 4.4.
Research Insight content	**Point:** Acquisition experience can help firms to extract generalizable routines and lessons learnt from previous acquisitions. Anyway, firms need to employ deliberate learning mechanisms. **Counterpoint:** Experience can have negative effects on acquisition performance, including the risk of drawing false conclusions, overstretching correct ones beyond where they are applicable, or getting stuck in organizational routines that fail to adapt to changing circumstances. **Contingency:** There is no significant general effect of experience, and acquirers should be aware of more complex interactions with other variables. There is evidence that transferring experience in deliberate planning can help firms to achieve the desired outcome faster, especially when employees play an important role. While managers should avoid intuitive decisions as the transparency decreases, the transfer and sharing of resources and capabilities often requires intuitive decisions that are more flexible and faster to achieve, with quicker synergy realization.

Table 6.17

References	
	Point: Zollo, M., & Singh, H. (2004) Deliberate learning in corporate acquisitions: Post-acquisition strategies and integration capability in U.S. bank mergers. *Strategic Management Journal*, 25: 1233–1256. **Counterpoint:** Heimeriks, K.H., Schijven, M., & Gates, S. (2012) Manifestations of higher-order routines: The underlying mechanisms of deliberate learning in the context of post-acquisition integration. *Academy of Management Journal*, 55(3): 703–726. **Contingency:** King, D.R., Dalton, D.R., Daily, C.M., & Covin, J.G. (2004) Meta-analyses of post-acquisition performance: Indications of unidentified moderators. *Strategic Management Journal*, 25(2): 187–200. Uzelac, B., Bauer, F., Matzler, K., & Waschak, M. (2016) The moderating effects of decision-making preferences on M&A integration speed and performance. *International Journal of Human Resource Management*, 27(20): 2436–2460.

Notes

1. Osterwalder, A., Pigneur, Y., Bernarda, G., Smith, A., & Papadakos, T. (2014) *Value Proposition Design: How to create products and services customers want*, Wiley, Hoboken, NJ.
2. RB does not disclose individual gross or operating margins on acquired brands/companies or divisional results, hence the focus on averages.

7 Next Steps

If you ever find yourself wondering when the real M&A negotiations will take place, it is already too late. They probably began a long time ago.

A lot of people think that the negotiations only begin when you call the seller. Applied deal tactics and negotiation skills start long before this 'first point of contact,' because whether you realize it or not, your company already has a reputation of some kind in M&A. Maybe you are known for not doing any M&A whatsoever. Maybe you are known for your M&A failures. Or maybe you have a reputation for being M&A savvy, with a brutal takeover approach and severe cost cuttings.

M&A Reputation as a Tool for Success?

Just over 10 years ago, one of our M&A Elite companies was not highly active in M&A. It was formally decided internally that there would be a new corporate strategy to create growth by acquiring competing firms. As it turned out, executing this new M&A strategy was not easy.

The firm felt that the international M&A scene didn't take them seriously. This made the CEO decide to hire a strong external advisor—an investment bank.

In the aftermath of its first corporate M&A transaction, the CEO realized that the internal corporate team were actually the ones to identify realistic cost synergies, not the investment bank.

After the transaction, the CEO felt that the price he paid for the investment bank, and their terms and conditions, were way above the value that was provided to his firm. So, before the next deal took place, he created a company-specific M&A engagement letter with heavily revised pricing, terms, and conditions.

Nevertheless, the investment bank did deliver one single and brilliant idea—they coincidentally knew that some of the existing shareholders of the target company were unsatisfied with the current management and their results. When presented with the operational efficiency offered by the acquiring company, they indicated that they would be willing to back a bid. This enabled the CEO to approach the target company, and a meeting was arranged by the investment bank.

During this first meeting, the CEO presented the idea of a complete takeover (in a diplomatic way) to the target CEO and the board. They were not interested at all. However, when he presented the backing of their investors in a (very) close-to-majority shareholding position, the target CEO felt as if he was on a burning platform.

Imagine being left with the choice of accepting a bid with a serious premium to your stock price backed by your biggest investors, or turning it down and facing their wrath once discussions became public. It reminded us of US President Roosevelt's foreign policy motto: "speak softly, and carry a big stick."

In future deals, this company followed the same business model-driven M&A approach, driving hard on the right building blocks and achieving cost synergies. An M&A reputation was suddenly born.

Can You Build a Good M&A Reputation and Attract Sellers?

The ultimate goal for any would-be corporate acquirer is to have the sellers come to them. Ideally, these sellers would be part of a family-owned entity, caring very much for the future of the business and passing over a strong legacy.

Some companies underestimate the value of attracting a quality seller, and over the past decade or so there has been a trend towards financial sponsors (such as private equity firms) winning deals over industrial buyers. This is an unusual occurrence for two reasons.

Firstly, industrial buyers would normally have higher deal synergies and so be able to outbid the financial sponsors. However, the prevailing, historically low interest rate levels and abundance of capital have reduced this gap, and financial sponsors have shown an incredible drive for optimizing existing corporate businesses, which competes with traditional M&A business model drivers.

Secondly, financial sponsors are outrageously good at building their own hype. The reality is that many corporate sellers enjoy life with a financial sponsor, and they are happy to share their positive experience with other potential sellers. One of our M&A Elite companies has taken inspiration from this approach and uses its previous acquisitions to act as speaking partners for new potential sellers. Family-owned entities like to speak to other family-owned entities about their journey—they understand each other.

Of course, your company itself should also act as a cheerleader for your M&A success. However, it is important to know when and how to bring certain departments into the mix.

Certain functions—like legal counsel, corporate strategy, or business development—work well alongside M&A, but it is not always necessary to have all the other departments involved. In some cases, it is an outright disadvantage to have marketing and sales people from the (potential) acquiring company involved. There is no transaction before the closing of the deal—both parties remain competitors until the deal has been finalized, approved, and fully completed.

This is often a controversial matter in corporate M&A. On the one hand, you want to involve as much corporate knowledge as possible. On the other hand, you want to control the process and make sure you have a total overview at all times.

We believe that a successful acquirer should put the most emphasis on controlling the process and sometimes even choosing to establish a 'clean team' without any sales or marketing people.

It must be crystal clear that the M&A department is in charge and that no-one starts a dialogue or a relationship with a target company without being 100% coordinated with the M&A team.

In our discussions with M&A professionals, we have often heard how inconvenient it can be to become involved in a deal when the first initiative has been taken by a business unit that is not fully aligned with your M&A Formula and Launchpad.

Particular damage can be caused when these entities don't pay attention to the competition authorities, which can cause additional adverse consequences, particularly for the acquiring company. Even smaller mistakes can still lead to nasty inquiries and long disclosure processes. One poorly worded email or comment could be misunderstood or misinterpreted as the parties having coordinated commercial activities or future corporate strategies.

Things you SHOULD NOT discuss during the early stages of M&A

- Clients (particularly future coverage, allocation, focus, tenders/bids)
- Channels (marketing, route-to-market initiatives and changes)
- Key activities (R&D is not yet widely known)

- Revenues (discounts, reimbursement policy)
- Partners (long-term contracts, terms and conditions)

Things you SHOULD do during the early stages of M&A

- Involve your entire organization in your 'should be' business model
- Sign the appropriate confidentiality agreements

As we have seen with the M&A Deal Committee, there are a lot of ways to involve the broader organization in M&A transactions. Most of your corporate finance neurons are actually in the business units themselves, but discussions with targets should be limited to company presentations and perhaps a positive comment on how you will be able to 'do great things together,' but no more than that unless the business unit has been trained to do corporate M&A.

Any discussion must be secured by signing an appropriate confidentiality agreement, which determines who receives the information, what data is to be used, what happens if the transaction is dismissed, and any other points of consideration. It should also confirm that any information is used solely in the evaluation of this specific M&A transaction.

If strictly confidential information is needed in the valuation of business model fit, then make this happen after signing an LOI. Our M&A Elite have a much wider understanding of the transaction at the LOI stage than we normally see in other companies. Many companies, to our surprise, will celebrate any kind of signature on the most weak, non-binding LOI, as if there was a deal on the way. Some companies even keep the nasty parts until later in the negotiations, which we believe is outright wrong. Such issues will only grow larger when not mentioned upfront, and will take up a lot of resources on something that was never going to materialize in practice.

This is an example of a good failure—fail fast. ASSA ABLOY is no stranger to simply walking out of meetings once it becomes clear that the deal is not a good business model fit or if the seller does not seem genuinely interested in staying long term and create value together with ASSA ABLOY. No company can justify spending any extra time, money, or energy on a deal that isn't going to work. Better to get all the issues out in the open right from the start, so you can fail as quickly and deliberately as possible and move on.

Things to Avoid Internally—Bias

What you don't know about bias can hurt you as a business owner and could work against shareholder value interest. Remove your deal-destroying assumptions and stick to your business model-driven M&A approach. Strictly speaking, an M&A Deal Committee can also be biased, but the fact that it consists of members from the whole organization reduces the risk of bias from individuals with personal agendas.

There are a few ways in which bias can affect a deal.

Selection Bias

When M&A is run by department heads from a business unit, you will more often than not have a principal–agent theory problem. The shareholders do not get a higher total return because some department heads think it would be cool if they could buy a big company in a specific country and thereby elevate themselves to head-of-region globally in the firm. Do regional bosses really think like this? Yes. This happens all the time and it's what we call 'M&A pet-projects.' The managers start to get emotionally involved, because a specific deal can do something for them personally. Beware of selection bias deals, which are at best non-value-adding to shareholders.

A similar scenario could also play out in what is identified as 'the perfect merger.' Assume your finance director (let's call him Sean— 59 years old, been with the company for 30 years, looking forward to his last 4–5 years before retirement) is running financial due diligence.

The other firm also has a finance director (let's call her Lucy—34 years old, highly driven, with an impressive academic record and professional resume). Sean might feel a slight pressure here, he may have a selection bias to not merge the two entities because he is afraid of losing his position, compromising his retirement plans, or being forced to make huge changes under the new regime. While you might feel empathy for his situation, the shareholders get absolutely no value out of his selection bias problem.

Confirmation Bias

This is the attitude that says 'it has always worked for us.' Sometimes a firm will decide that because a specific type of acquisition has worked out for them in the past, it will work out for them again in the future. But this could easily become a confirmation bias problem, and such issues could also have an embedded principal–agent theory problem as well. From a business model-driven point of view, it may be smart to repeat past deals (for instance, where your Business Model Canvas building block is 'Channels'). However, if you start believing in one solution and one solution only, you are confirmation biased. A Goldman Gates Scoring may suggest it's the right building block, but the M&A complementarity (as seen in 'The Antecedents of M&A Success') will be very low, which means you must handle things with (extreme) care.

You can just as easily have a negative confirmation bias as a positive one. Either way, it is not a route to M&A success; both selection and confirmation bias can work against M&A success, whether you buy/merge or not.

Cognitive Bias

This last example of a potentially negative bias is probably one of the most important reasons for Danaher's success. From what we have seen, Danaher tends to look for not-so-fit companies which it can turn into very-fit companies. Danaher is essentially a corporate fitness organization, which finds members who are not fit and turns them into athletes. Why don't these corporates just do it for themselves, when Danaher does it so easily? Even with companies that have a very low complementarity score in the M&A Goldman Gates! This is because of cognitive bias. Too many firms get stuck in a rut, where they budget for a certain annual increase in taxes, raw materials, etc., without being able to match these increases with their own productivity. They are not prepared for change.

Danaher relies on an ecosystem in which there are many corporates with a cognitive bias. Danaher knows that it can convert these companies, with or without their management, and so they will. If you have cognitive bias, you may lose some good deals because your own firm is not fit or because you don't believe you can make another fit. Otherwise, a Danaher-like firm will just have your business for lunch someday. This also adds shareholder value, so long as the prevailing shareholders get a premium that is commensurate with the value of the new, lean firm.

Where to Get External Advice?

Implementing business model-driven M&A takes guts. But if you've come this far, you will already know that it works.

The next challenge is to stay on target. You've put in all the hard work of building your M&A Launchpad, following the M&A Formula, and creating your anti-portfolio. You've seen M&A success and grown your business. Why rest on your laurels now?

This is the perfect time to build your resource library and surround yourself (and your company) with all the tools to ensure that you are on track to join the M&A Elite.

FON-A-PAL—Building Your Resource Library

One of the most undervalued resources at your disposal is your peer network, which we refer to as 'FON-A-PAL.' Social media is a great way to stay in touch with colleagues past and present, and to stay on top of the latest developments in your sector. But you can really broaden your horizons by attending conferences, talks, and courses, where you can meet new people with a variety of backgrounds and opinions within M&A.

Across the international M&A community there are many interesting peer networks, such as HR, procurement, treasury, sustainability, and so forth. What is interesting about these peer networks is that you are learning directly from your peers. And on top of that, you easily build up a network with people who are asking the same questions as you, and others who might know the answers. For instance, if you are pursuing a deal in Myanmar, you can simply 'FON-A-PAL' (call your M&A friend and ask: "Are you guys doing anything in Myanmar? What WACC do you use? Who should we use for due diligence? Know any good local lawyers?" You will get some of the best advice on the planet, and it's all for free!

There is enormous value to be found in a simple conversation—as long as you are open to listening and learning at every opportunity. This goes both ways. If you're really respected among your peers, you can expect a few phone calls yourself.

These people are just like you, and by building up your peer network, you are creating an incredibly strong and trustworthy resource.

According to Edelman's 2016 trust barometer, the financial services are among the least trustworthy sectors in the world, while technology firms are scoring among the highest.

Edelman suggested that these results were down to the fact that more and more people now tend to trust people like themselves—financial services are seen as a distant, almost threatening presence, while technology firms are on the front line, connecting people from around the world.

If you are an expert in your field, 'a person like yourself' might be a technical expert in another discipline (e.g. a legal advisor) or a top academic. Or maybe it is an M&A colleague with similar responsibilities.

Once you start to put yourself out there, you will find that your professional network grows quite organically.

These are the people who will challenge you, support you, and help you to take the next steps in your career.

Never expect an external advisor to bring you M&A success. They might make you feel as though they are 100% on your side, and of course they prefer you to be doing good and solid M&A deals, but when it comes down to it, they don't really care what happens to you or your business. They are highly focused on getting their 'success fees,' and they will get them either way—even if the deal fails. It is perhaps about time you start testing things like this in your 'FON-A-PAL' network. Did you ever hear from an investment bank that you should use a high-risk adjustment (WACC) in your M&A project? Nope, because that decreases the chance of pocketing the success fee. This is a question we always ask peers in our network.

Having said that, external advisors do have a place in the M&A Formula. However, they are suppliers and should just be seen as a helping hand rather than a lifeline.

What Do We Mean When We Talk About External Advisors?

External or 'sell-side' advisors are any company which sells services and/or products to corporations during the M&A process. These include:

- Financial advisors like traditional banks or investment banks with a distinct focus on corporate finance or M&A.

- Legal advisors and law firms.
- Accounting firms who offer so-called transaction services in relation to M&A. While this mostly refers to due diligence, some big accounting firms have actually taken up advisory services in recent years and find themselves competing with the investment banks and financial advisors.
- Corporate diagnostics. A highly reputable firm like McKinsey can become an outstanding external advisor in M&A as it is not biased towards making you do an M&A deal—it is simply a firm to consult when you consider 'go'/'no go' on a deal. McKinsey is a balanced advisor, which is not afraid to present you with stretch goals, or tell you when they believe you are not on a realistic mission. McKinsey's only mission is that they are successful if their clients are successful. McKinsey is highly analytically driven and systematic.

There are other very balanced advisors out there who will not be biased when it comes to discussing your existing business model and potential add-ons via M&A. One of the companies I have worked with is Business Models Inc. It is among the best in the field when you want sparring on your existing business model. It is always worth asking these advisors: "Is M&A really our best alternative?" A good start for the M&A Formula is to be aligned within your company on your existing business model. Where are we now? What are we good at? How do we know?

Financial Advisors/Investment Banking

A financial advisor has to comply with highly strict onboarding rules when giving advice to private individuals about investing in financial assets. The so-called know-your-client (KYC) directive must be signed proper by both the bank and the investor. The investor will be asked about investment horizons, tax, risk willingness, asset allocation between various assets, currency risk, and so on. The financial advisor needs a special background and education to comply with all the regulation and diligently inform the investor about the possibilities—particularly how they can lose money and perhaps even risk some assets being illiquid if they are not stock listed. This is why the absolute majority of all investments to private individuals are held in highly liquid assets and the world has increasingly turned to so-called cheap beta products, because no one really believes that financial advisors are better than the market return.

If the investor is an SME firm looking to buy a company, there are no rules. In practice, that same SME owner is often also a private investor. When she invests her private wealth, there is an abundance of rules and regulations taking care of her, but as soon as she does corporate M&A in her firm, there are no rules whatsoever to protect her. She can afford to make mistakes and losses with her own personal wealth, but once the business is involved, there will be clients, lenders, and employees to consider—yet she can make these big company decisions without any regulation or professional oversight. That is strange, isn't it? As with investments, there are risks involved in any M&A transaction. For instance, you are on the hook after the investment contrary to stock-listed equities, which can be sold and exchanged for cash within three days.

If you are an SME owner by day and corporate M&A superhero by night, you have nothing to worry about. But if not, you may need to draft in some professional help at some point from a financial advisor or investment bank.

We are about to do some serious myth killing, not only about M&A high rollers but also SME advisors. You will find that this book's vision of lowering the M&A failure rate actually has very little to do with whatever external advisor you choose for financial matters. When a deal fails, you still have to pay your advisors. Even if you get bad advice, they will usually be protected by an indemnity clause in their contract. Your advisors will not bear the brunt of your M&A failures—that is entirely down to your company.

The size of the company doesn't matter, nor does the size of the deal. The CEO needs to do her homework and realize that no external advisor can offer any remote guarantee of lowering the M&A failure rate. You can only do this in your own company.

Think about it. Most investment banks charge a 'success fee' of between 0.3% and 3% of the enterprise value of the company involved in the M&A deal, but how do they define the word 'success'? Very loosely. To them, a successful deal is simply any deal that takes place, regardless of whether it falls through.

We find it very strange that people still sign such one-sided engagement letters with external advisors, no matter whether it's Goldman Sachs or the local bank next door. External advice is not commensurate with M&A success. As far as we're concerned, that is just articulate routine work for an M&A professional or what we call 'hygiene factors' in M&A.

This book's mission is to improve your M&A success rate, and that has nothing to do with the external advisor. It is your responsibility as corporate acquirer or partner, who will merge with another company.

The tipping point for you or your organization is when you have your moment of transformation and decide to do something about the problem of M&A failure.

The moment of transformation is key—when a lot of people, and perhaps even your own subconscious, start telling you it cannot be done. You cannot possibly create corporate M&A success when the outcome globally is less promising than flipping a coin. The moment of transformation is the single most important step in personal success—the moment when you decide to make a change.

Our mission with this book is to make you believe that you can do it. You too can have corporate M&A success over and over again.

The M&A Formula is your non-bullshit system and it's been proven to work. It is a reliable tool which is easy to use, and once you've had your moment of transformation you can start right away.

We want this book to give you self-confidence in M&A. The whole world of corporate M&A has a 60–80% failure rate, some would even claim it's as high as 70–90% failure rate, but that doesn't apply to every company or any situation. However, this statistic makes people nervous, and means that companies believe they are better off working with external advisors. But change must come from within. Remember, the more M&A deals you do, the more money the external advisors will make and there is no evidence in the world that external advisors can turn your M&A activities into success. It is time for you to do your own homework.

Some companies can't see the wood for the trees. They might rush to hire the most expensive and highest ranked (by size and numbers of M&A deals) financial advisors who have proved, over more than 20–30 years, that the outcome is definitely not M&A success for their client but in reality the definition of insanity—doing the same thing over and over again expecting a different outcome. But sometimes the obvious solutions are a little too obvious for us to see. Most organizations actually have the human resources readily available for M&A success. Make your organization punch above its weight—invite various disciplines. Form an M&A Deal Committee. Use the M&A Formula. It is easier than you think.

When an M&A deal does not deliver on its promises, there are the instant repercussions and the slow-burners. For instance, in the immediate aftermath of a failed merger or acquisition, a stock-listed firm will generally see a drop in its share price; media coverage will be less than complimentary; and a couple of M&A executives may leave. In an SME firm, you could lose money and your long-term reputation could take a hit.

But that's only the beginning. When the next quarterly figures come in, the cost of the failed deal will become clear and questions will be raised all over again. Then there is the next shareholder meeting, where you can be sure that angry investors will not hold back. Needless to say, each of these events will garner even more media interest and will inevitably affect the share price. Furthermore, it can be hard to prepare for the actual fallout, as shareholders have become much more vocal in recent years.

Ultimately, the long-term repercussions will be shouldered by the C-suite management, no matter how involved they were in the deal. It is the CEO, the CFO, the COO, and the CRO who will have to answer the tough questions and account for this failure. In a worst-case scenario, if a governance issue is thought to be behind the failed deal, one (or more) of the C-suite could be booted out, fined, or even prosecuted. In an SME, it's even worse because it is not only a job but a way of life. The wrong partner, in business as in life, could mean misery.

That is why it is so important that C-suite executives are involved in every process of every M&A deal, every time. The reasons for the deal must be transparent to everyone involved, and the business-owner within the organization must take responsibility for each case. There is simply no excuse for complacency, especially when it comes to business model-driven M&A, where every building block is turned around a few times to ensure that the whole organization is punching at its full weight.

Deals do fail. A lot. And failure costs money and damages corporate reputations. No matter how efficient your M&A department may be, the figures speak for themselves. You must have a mission statement if you want to avoid being the next M&A failure.

So How Do You Protect Yourself Against the Fallout from these Bad Deals?

Well, first of all, you need to reduce your corporate failure rate. In a perfect world there would be no bad deals for you to worry about, no profit loss and no damage control. But the only way to get around this is by making the CEO design a bespoke business model-driven M&A plan. To be blunt, the CEO should do more homework.

Ask your team to talk you through every single building block in the business model and get your senior executives to advise you before you sign off on anything, so that they too are involved in the process.

Next, you should make sure that you understand every aspect of the deal being suggested. Ask your M&A team tough questions: What would happen if we didn't proceed with this deal? How much will it cost us if the deal falls through? What regulatory issues might we face on a regional/national/international scale? How did you come to the conclusion that we should go ahead with this deal?

Treat Every Single Deal as Though Your Career Depends on it, Because it Just Might!

CEOs must get back to where it all started. What is my (new) business model? The original Business Model Canvas was created to help and motivate lean start-ups and it appears to work brilliantly for M&A deals. It's about knowing your customer. It's about designing a new business model for your customer. Understand it and take charge.

Finally, don't rush into anything. You will be pressurized on all sides to sign off on deals quickly, under the guise of 'saving money' or beating your competitors. Don't sign anything until you are confident that you understand why you are doing the deal, how much money it will cost your company, and what it will add.

Jens Bjorn Andersen and Jens Lund, CEO and CFO of DSV, worked with their team to make one noteworthy achievement. Despite numerous opportunities (not to mention pressure from investors), the company did not succumb to the temptation to do a deal just for the sake of it with the exception of acquiring a rather small Norwegian firm in 2013—in fact a loss making division in 2014 by which the team learned that it was not easy to execute on smaller M&A deals.

In fact, the firm went almost 7 years without doing a single M&A deal, before acquiring a US firm in 2016. The largest acquisition in the history of DSV. Newspapers will never report on this kind of achievement—they only want to talk about the deals that have been done. When an ongoing deal does not materialize, they call it a failure. Don't waste your time worrying about the media. It has absolutely nothing to do with an M&A Formula followed by M&A success.

When deals go bad, everyone will want a scapegoat, and the boss is always first in the firing line. You have to protect yourself from the outset, so you don't end up paying for someone else's mistakes. Find someone who will challenge you, because very few people actually will if you are the boss!

In rare cases, you may actually benefit from an opposing team whose mission is to tell you why you shouldn't do an M&A deal. How often does that happen when all your advisors are being paid to make deals, not stop them? Be smart, be vigilant, and don't be afraid to say no if the deal is not commensurate with your own M&A Formula.

M&A Myth-Busting

There is a popular stereotype that the perfect M&A deal goes something like this:

- Hire a top investment bank.
- Perform a company valuation.
- Lay out the legal framework and documents.

But even for firms who follow this project plan to the letter, the M&A failure rate is still frighteningly high.

The perception used to be that corporates which surrounded themselves with external, highly paid advisors would fare better than others. However, we have found no evidence that this lower the failure rate and even less so is your road to M&A success.

In fact, a growing body of evidence indicates that the key to M&A success lies in internal transparency and crystal clear communication within both organizations. We have shown you examples of this from the M&A Elite and the world of academia, with SMEs.

This new phenomenon is set to radically drive down the corporate failure rate in the future, and eliminate the following popular (and expensive) M&A myths.

Myth No. 1: Hire an Investment Bank and You'll Be Fine

It's easy to assume that hiring a leading group of financial advisors will have a positive effect on your own company's M&A activities. However, there is absolutely no evidence to suggest that hiring an investment bank will lower your corporate failure rate. That doesn't mean you shouldn't hire financial advisors—in fact, we believe you should. Just remember that they have a vested interest in your firm doing the deal, no matter what.

Instead of relying on expensive investment bankers, make your own M&A engagement letter, aligned with your internal rules and procedures. Signing a traditional engagement letter can lead to serious cash drains for your company, even if you never do the deal! It is also worth remembering that in the event of the bank making a miscalculation, you will most likely have no financial coverage at all, as the liability clause requires the bank to have been grossly negligent in these cases.

Myth No. 2: Valuation Is Paramount

To be brutally honest, valuation is a pointless (in the future) free cash flow, replicating a perpetual bond. Depending on your geographic location, your deal will be subject to a discount rate, but no matter where you are buying, the investment bank will generally have a lower WACC compared to the corporate.

Why do you think that is? Every time you meet some head of M&A (corporate buy-side), ask them: "Did your bank ever tell you to increase the WACC in a project?" You will struggle to find one single head of M&A who said yes. By keeping the WACC as low as possible, as sell-side advisor you increase the purchase price of the company. It's just an approximation for the non-linear correlation between yield and price (or in M&A, WACC and enterprise value [EV]). There are so many unrealistic assumptions surrounding WACC and EV that we always mention manipulation when talking about valuation. EV assumes all cash flows are reinvested at the WACC discount rate. Whatever timespan you invest in, the same yield applies for one day, 10 years, or to infinity. Well, we all make these calculations as a supplement, but many advisors make it a big thing. It is better to use many other pricing mechanisms to support these calculations and never take the external advice on valuation alone.

Myth No. 3: Getting Access to New Markets and Products Is the Whole Point of M&A

No, it is not. There are several different types of deals in corporate M&A and when you use The M&A Formula you will understand the power of Businessmodel Driven M&A. There are so many other interesting deals and most of them come with a much higher chance of a successful outcome and within a shorter time frame than those mentioned here. The highest failure rates are more often seen when it comes to product extensions, cross-selling, and new markets. Furthermore, what one thing we have noticed from the Global M&A Elite is that they rarely count in expected revenue synergies as part of the valuation.

Myth No. 4: Getting the Synergies Right Is Half of the Battle

Missed synergies are either related to bad integration post-deal or inflated expectations by the C-suite. Or so they say.

Getting the synergies right in an M&A deal is not about understanding the new business model. It is about getting the business model to work correctly. A business model-driven M&A approach starts with your value proposition and your customers. Why would they buy your products or services? Too many corporate buyers rely on the spreadsheet and do not take the time to really understand what they want the business model to be in their M&A transaction, which is the foundation of any deal.

As the boss, you have to take charge of your M&A activity. Forget everything you thought you knew, and reduce your failure rate with your own bespoke M&A Formula.

It is definitely still a good idea to work with organizations that have a good grip on the operational side of M&A. Just remember, it is your company—whether you are a FTSE 100 firm or a novice SME—that will ultimately bear sole responsibility for your M&A failure rate.

What Your Success Looks Like?

"The first time I really understood what a completed successful M&A deal looked like was when I left sell-side M&A to work for a former client (FrieslandCampina)," says Peter. "Coming over to buy-side M&A was a complete game changer for me. Although I was more or less doing exactly the same kind of work, the main difference was that the eventual success or failure of the deal would now be down to my bosses, and they would definitely—and justifiably—blame me. I could not just pocket another success fee and move on to the next deal. Any deal could now ruin my hard-won reputation and the reputation of the firm."

"This may sound scary, but in fact it was highly rewarding. I would meet with our shareholders at least once a month; and at

FrieslandCampina, these shareholders were exclusively farmers. I wanted to make a good impression at the first meeting so I drew on my own experience of growing up on a farm when I made my first introduction."

"I knew from personal experience that when you are getting out of bed at the crack of dawn to milk a cow, you just hope that the milk is being sold for the highest possible price and utilized as much as possible."

"The first time I met with these farmers, I presented the simple idea that no M&A deal should ever be considered if the farmers could not expect a higher milk price. I was told by the external accountants how complicated this calculation would be. However, by working with the Finance department and members of the M&A Deal Committee, we were able to quickly calculate each M&A deal's expected impact on the milk price. Now we could all agree on why we were doing M&A in the company, and everyone was on board and fully aligned."

Calculating the Impact of M&A on the Price of Milk

- One cow in Western Europe yields around 25 liters of milk per day.
- Q is your quantity, which equals 25 liters.
- P is your price.
- What you get paid is $Q \times P$.

    Any business can be defined in a simple equation—there's no need to make it complicated.

What Does M&A Success Look Like?

Before any M&A success, there must be a strong alignment between the shareholders (the principal) and the management (the agent). This is another common thing we have noticed in analyzing the M&A Elite. There is a strong alignment between M&A activities and the way shareholders and management are being paid. For RB this is possibly the single most important driver for M&A success.

In a stock-listed firm, value must be created for shareholders when executing business model-driven M&A, otherwise there is simply no point in doing corporate M&A.

"When I hear a company arguing that a stronger presence in a particular country or region would create a stronghold or close the gap to competitors, I always ask myself what the shareholders are getting out of it," says Peter. "I will think of the farmer milking his cow and wonder how much happier he would be if the next M&A deal increased his cash flow. And how indifferent he would be to know that the market share was 28% or 29% if it had no impact on his earnings."

However, even a high milk price or EPS may have its drawbacks. The focus must always be on the building blocks, not on EPS itself. You have to be able to explain to the organization why you do M&A (e.g. to create long term shareholder value) and what is expected from each member of the team.

What Is the Key to Success in Corporate M&A?

The most important thing in successful corporate M&A is YOU.

You might feel discouraged by the negative statistics and scare stories about M&A failures. But this doesn't need to be your experience. Once you make the decision to transform your business through M&A, you have reached the tipping point.

Next Steps

You cannot apply the M&A Formula without first making the decision to transform your business with M&A success as your realistic goal. And implementing the M&A Launchpad *without* the formula would create additional adverse consequences, as you may reduce the number of 'bad deal failures,' only to realize that you have just increased the number of 'really bad deal failures.'

Before, repeated and successful M&A was the realm of the M&A Elite, but now, when you reach your tipping point, you can take ownership of corporate M&A, because we've given you the tools—the M&A Formula, the Launchpad, and maybe even a potential M&A role model to show you how to achieve M&A success yourself.

There Are Only Seven Companies in the M&A Elite; Is that Really It?

Of course not. The reason we chose seven companies was just to illustrate all building blocks in the Business Model Canvas. There are many more companies out there including hundreds of SME with M&A success (that's our next book by the way). However, we wanted to find the 'investor darlings'—the companies with an unusually good reputation among investors because of their acquisition records. These seven names came up again and again, and we were able to document their above-average total return to shareholders over a 10-year period. These returns were heavily impacted by hundreds of M&A deals. Add to that the hundreds of SME firms who have seen M&A success across even more deals, and a pattern starts to emerge …

If you are still not fully convinced, just remember that we have also shared with you new and trending M&A concepts, such as the M&A CFA, as well as other M&A accelerators, like functional M&A software to secure a CTP organization, which is crucial in M&A. Furthermore, the M&A Dashboard ensures that your colleagues/employees do not work against you in your future M&A deals. If they do, at least you will know before it's too late to react. You can even test your bespoke

M&A Formula in a real life simulator through gamification (see more on www.fixcorp.co)

None of these accelerators were in play just a few years ago, but they are available to you as soon as you have reached your tipping point.

You Too Can See this Success

Find the building blocks you are really good at. Leverage them via corporate M&A. Choose one or two M&A drivers and choose a few metrics for success. Apply as few metrics as possible and drive hard on those. This simple articulation of your goals will then help you with the second pillar in your M&A Formula: communicative leadership.

Everyone in your company should be able to answer the question: "Why do we do M&A?"

Leadership is a very powerful tool in the M&A Formula. It works from the very beginning, as it forces you to discuss the business model fit from the start. It is also much easier to be an effective leader when you have a solid mission statement (business model-driven M&A).

Celebrate when you fail fast in M&A discussions. It is only the investment bankers (who lose their success fee) and the newspapers (who love a dramatic headline) who will describe your fail-fast M&A deals as 'failures.' They are not failures, they are successes, because you just added another company to your M&A anti-portfolio. They are 'good' failures.

Take Ownership in M&A

What we have learned from working with both the M&A Elite and SMEs is the fact that the more ownership you take, the higher the chances of M&A success.

Corporates who take proprietorship of their own way of doing acquisitions and/or mergers become learning organizations. Make your own company-specific M&A Formula and get that M&A Launchpad ready. Excellent leaders give no orders, remember? When M&A becomes another day at the office, you start to become excellent.

What If You Only Want to Do One Deal—To Buy the Company Down the Road?

Just remember that any M&A sell side advisor make more money, the more deals you do. You will not necessarily hear from them if you are about to make an M&A failure. What they really care about is their fee and particularly those sell side advisor who even has a success fee, which has absolutely no relation whatsoever to you being successful in M&A.

We invite all financial advisors and investment banks to challenge us on this. Our statement is that none of them have a track record of M&A success, like the M&A Elite or the global M&A-savvy SMEs in our study. These advisors are highly useful as support in your M&A deals, just as long as you remember two things: firstly, they will not give you M&A success; and secondly, you need to contract them with a company-specific M&A CFA, which is best practice in corporate M&A today.

We are happy to stand corrected and pledge to publish any such material in the *Harvard Business Review*, the *Journal of Finance*, *MIT Sloan*, or any serious and trusted paper that might be interested in such an analysis.

We write all this to remind you—the reader—that your moment of transformation, followed by implementation of the M&A Formula, is the fast way to M&A success. By following this route, you and your company will become a learning organization that will learn from every single deal—even the ones you don't do, as they make you anti-fragile thanks to your M&A anti-portfolio.

You will gradually become leaner and better, and within a few transactions you will be an M&A champion. You just have to make the decision first.

May the M&A Formula be with you …

INDEX FOR CASE INSIGHTS & RESEARCH INSIGHTS

Page references followed by f indicate an illustrated figure or photograph; and page references followed by t indicate a table

A

ABLOY 188
ABX 136
Accusort Craftsman 168
Acqua di Parma 153
Acquisition experience 198t
acquisition, positive or negative effects of 164t
acquisitive growth 148t
Aerogard 141
Air Borne 141
Air Wick 141
Amphyl 141
antitrust authorities 160
Ardbeg 153
Arnault, Alexander 150
Arnault, Antoine 152
Arnault, Barnard 149, 150
ASSA 188
ASSA ABLOY 136, 186–98, 187t, 205
   acquisitions in 186–7, 191
   ASSA ABLOY Way 190–1
   Entrance Systems business 195
   Business Model-Driven M&A 198–9t

B

belonging, feeling of 181
Belvedere 153
Berlutti 153
Besam automatic door business 189
'Better Business' Business Model-Driven M&A 145
bias, in-group and out-group 185t
bolt-on acquisitions 189
Bonjela 141
brand awareness 161
brand redeployment 147t
brand value 158
brands 140–1, 145–51
   acquired 143
   international 149
   luxury 149
   premium 161
   quality 152
Brasso 141
Brisset, Matthieu 152
Buffet, Warren 158
'build and leverage' 189
business development 158

business model drivers 159t
business model fit 153, 190
Bvlgari 153

C
Calgon 141
Campina 175, 176, 179, 180, 181
cannibalization effects 148t
capability development 164t
CEDES sensors 194t
Céline 153
CEOs 150, 184, 201–2, 213, 215, 216
Cêpacol 141
Château d'Yquem 153
Chaumet 153
Christian Dior 152, 153
Chäteau Cheval Blanc 153
Cillit Bang 141
Clearasil 141
client base, broadening 144
client knowledge 145
Cloudy Bay 153
co-branding 151
communication 159
corporate takeover 149
cost as short term business model driver 184t
cost efficiency ratios 143
cost synergy 135–6, 143, 145, 160, 175, 178, 202
cost/income ratios 143
Crawford 194t
cross-domain experience transfers 164t
customer segments 155t

D
Danaher 34, 53, 58, 60f, 83, 84, 136, 165–7, 167t, 172–3t, 207–8
Danaher Business System (DBS) 166
d-CON 141
De Beers 153
Dettol 141
DFDS Dan Transport 136
DHL 133
Dom Pérignon 153
Dover 168
DSV 136
Durex 142

E
earnings before interest and taxes (EBIT) 188, 196
earn-out based on company profitability 190
Economic Value Added 158
Edun 153
Emilio Pucci 153
EPS 175
experience, prior 164t
export ratio 134

F
fast-moving consumer goods (FMCG) companies 160
Fendi 153
Finish 142
fixed hurdle rate 158
Fluke Networks 168
Frank's RedHot 142
Frans Maas 136, 137
Fred 153
French's 142

Friesland Foods 175, 176, 178, 180
FrieslandCampina 113–16, 220
'from push to pull' 178, 181

G

Gaviscon 142
Gendex 168
geographical market presence 190
Gilbarco 168
Givenchy 153
Glass Plus 142
Glenmorangie 153
Guerlain 153

H

Hach 168
Harpic 142
Harvard Business Review (HBR) 169
Hennessy 153
heritage 149–56
Hermes 150
historic deals 158
hostile takeover 149
Hublot 153

I

ibuprofen 144
incentives 147
in-domain experience transfers 164t
internal risks 185t
intuitive decisions 198t
investor community 157

K

KaVo 168
Kenzo 153

Kollmorgen 168
Krug 153
Kuehne & Nagel 133
K-Y 142

L

Lange Trojan 168
Larsen, Kurt 137
law of averages 186
L-Door 192t
lean system 169
Leica Microsystems 168
Lemsip 142
letter of intent (LOI) 93, 94, 95, 109–11
Lighthouse 192t
Loewe 153
Logistic and distribution network savings 196
logo 181
Loro Piana 153
Louis Vuitton 153
low-cost countries (LLCs) 196
LVMH 149–56
  brands and houses 153–4
  fashion & leather goods 153
  key resources and key activities 155–6t
  perfumes & cosmetics 153
  watches and jewelry 153
  wines & spirits 153
Lysol 142

M

M&A complementarity score 154f, 155t, 164t, 182t, 184t, 198t
M&A complementarity 159t

M&A Deal Types 159–61
   Type 1: Entering a New Market 159–60
   Type 2: Existing Market Position 160–1
   Type 3: Developing a New Company 161
   Type 4: Greenfield 161–2
M&A Drivers 57–8, 156–66, 181, 184
Magnan, August 166
managerial attention 185t
manufacturing footprint 196
manufacturing in low cost countries 196
Marc Jacobs 153
market exposure 175
marketing 148t
Matco 168
McKinsey 179, 211
Mercier 153
Moët Chandon 153
Moët Hennesy Louis Vuitton *see* LVMH
Mortein 142
Moynat 153
Mr. Sheen 142
Mucinex 142

N
Nassau 192t
Nergeco 194t
Nicholas Kirkwood 153
Nippon Express 133
Nurofen 142, 144

O
Office consolidation 196
operational autonomy 155t
operational excellence 134t
organic growth 142, 152, 156, 188
Osterwalder, Alexander: *Value Proposition Design* 141

P
Pelton & Crane 168
planning, meticulous 158
Portescap 168
post-takeover 152
product design 196
product revitalization 147
professional investor 143
profitability 190, 191

R
Radiometer 168
rainbow logo 170
raw materials 152
RB 141–8, 142t
   acquired Boots Healthcare Division 145
   brands 141, 144
   business model-driven M&A 145–6
   cost efficiency ratios 143
   timeline of acquisitions 145
re-branding 151
Reckitt Benckiser *see* RB
redundancies 155t
resources, transfer of complementing 139t

responsibility
　for acquisitive growth 190
　in M&A 190
return on corporate equity (RoCE) 190
Return on Net Assets (RONA) 158
Rimowa 153
Robinson, C.H. 133
Route 2020 strategy 181
route to market 158
Ruinart 153

S
Samson 136
Sani Flush 142
scale economies 160, 179
Schenker 133
Scholl 142
Sephora 153
shareholder value 136, 147, 206, 208
Sinotrans 133
speed in integration 140t
Starboard Cruise 153
Strepsils 141, 142
subsidiary consolidation 196
superstitious firms 164t

T
Tag Heuer 153
take out capacity 160
teamwork, importance of 166–8
technology 196
third party logistics (3PL) 133–9

Thomas Pink 153
Thomas, Patrick 150
transparency 159, 170, 182, 198t
Tullberg 134

U
unsolicited takeover 149
UPS 133
UTI Worldwide 136

V
value 172t
　long-term 157
value added growth 149
value chain 172t
value proposition 141–8, 147–8t, 151–2, 153, 162, 184–5, 191, 194, 220
Vanish 142
Veeder-Root 168
Veet 142
Veuve Clicquot 153
Videojet 168

W
WACC 158
Woolite 142

Z
Zenith 153

INDEX

Page references followed by f indicate an illustrated figure or photograph; and page references followed by t indicate a table

A
ABLOY 188
ABX 136
academics 6, 11, 13, 16–17, 25–6
accelerating the M&A Formula 117–28
accountability 77
acquired growth 86
acquisition launchpad 94, 104f
acquisitions as corporate building blocks 189
Adams Respiratory Therapeutics 145
Administration Skills 56
ADT 120
Aftersales Service 55
agile funding 95
Al Noor 106–7
Allen & Overy 99
Amazon 33
analytically driven processes 9
Andersen, Jens Bjorn 216
antecedents of M&A Success 17, 51–3
anti-portfolio 37, 58, 67–8, 77, 208, 224, 226
Anxiety Theory (Adams) 126–7
Apple 15, 33, 34, 35, 54
applied deal tactics 201

Arla Foods 179
Arnault, Alexandre 150
Arnault, Antoine 152
Arnault, Bernard 149, 150, 152
as is business model 40, 55, 149
ASSA 188
ASSA ABLOY 13–14, 22, 25, 34, 56, 71, 81, 85–6, 136, 186–98, 187t, 205
  acquisitions in 187–9, 191
  ASSA ABLOY Way 190–1
  Entrance Systems business 195
attracting sellers 203–6
Avago Technologies Ltd/LSI Corporation 101
AvidXchange 120

B
bad business model fit 12
bad deals, protection against 215–6
'bad' failure 91, 93, 94
bad M&A, formula for 12
bad-ish failure 46, 47
banks
  contender 115
  domestic 115–16

Index

investment 85, 108, 115–16, 202, 211–15
bargain price M&A 66
Barsballe, Jan xi, 25
Bauer, Dr Florian xi, 17, 25, 27, 58
beauty parade 111
Belfast 69
belonging, sense of 73
Berry Petroleum Company 101
Besam automatic door business 189
best practice in corporate M&A 99, 100, 102–3, 225
bias 87, 119, 170, 206–8
binding offer 94, 95
Biogen 34
BMW 54
bottom up approach 9
brand identity 56
brand recognition 56
branding 104, 141, 145, 153
Brexit 47
Buffett, Warren 158
'build' and 'buy' corporates 33
'build or buy' 35–6
building blocks 20, 25, 28, 39, 42, 45, 54, 55, 57, 78–9, 83, 158, 168, 188, 189, 207
business beat 127–9
business development 76, 104
business model canvas 25, 40, 41–3, 51, 53–4, 56, 57
business model drivers 2, 6, 26, 36, 38, 39, 40f, 41–4, 48, 63
   Geographical Fit 84
   Value Proposition 84
   M&A Value Drivers 84
   Market Position 84
   Overall Business Model Fit 84
business model fit 12, 45, 63, 65, 66–8, 77, 80, 84, 91, 109, 119, 141
business model-driven M&A 2, 6, 11, 17, 19, 24, 28, 36, 51–68, 72, 94, 118–19, 145
   following 38–9
   understanding of 32, 40–6
business model drivers in M&A 141–8, 151t, 167t, 172t
Business Models Inc. 211
business theories applied to M&A 126f
business units 77–8, 80
buying growth 44
buying with your head 45
buying with your heart 45
buy-side 1, 47, 86, 87, 93–4, 96, 102, 109, 111
Bvlgari store 150

C

Campina 57, 175, 176, 178, 179, 181
Capex 161
Carroll, Lewis: *Alice in Wonderland* 41
CEO 1, 4, 6, 7, 31, 35, 45, 78, 87, 89, 116, 150, 184, 201–2, 213, 215, 216
Cerner 34
CFO 31, 116, 215
channels 191
cheap beta products 212
Chicago Board Options Exchange (CBOE) 14–15
child labor 110
Cigna 34

Citywire 1000 'World's Top Fund Managers' list 7, 8, 9
'classical completion' accounts 110–11
client segment 46, 55, 145, 151, 152
codifying documents 103
cognitive bias 207–8
collaborative technology platforms (CTPs) 74
Comcast 34
common goal 77
communication 2, 6, 20, 38, 39, 73, 96
  bilateral 74
  leadership and 69–89, 119–20
communicative leadership 69–89
competition 53, 54, 98
competitors 57
  behavior, studying 9
complacency 215
complementarity 51–7, 59f, 207
  score 57–8
compliance functions 92, 94, 109
compound annual growth rate (CAGR) 170
confirmation bias 207
contingent liabilities 100
COO 215
corporate blah-blah 25, 63, 66
corporate bonds 8, 157
corporate diagnostics 40, 211
Corporate Framework Agreement (CFA) 95, 96–103
corporate lawyers 105
corporate social responsibility (CSR) 105, 110
corporate strategy 25, 38, 73, 113, 181, 182, 201, 203
cost efficiency ratios 143

cost synergy 55, 56, 57, 65, 85, 135–6, 143, 145, 160, 175, 177–84, 202
CRO 215
cross checking 51, 93
cross-border deals 110
C-suite 5, 70, 73, 77–80, 81, 99, 215, 219
'customer intimacy' 116
customer perception 79
customer relationships 55, 191
customer retention 79
customer segments 55
CTP 223

D

Daily Disposal 118
Danaher 34, 53, 58, 60f, 83, 84, 136, 165–8, 167t, 170–3t, 207–8
Danaher Business System (DBS) approach 166
Danske Bank 173, 174
data analytics 105
debt, net 111
decision gates 95
decision making 26, 45, 62, 92, 95, 120, 127, 189
  buying 36–8
delegation of M&A responsibilities 78
Dell Inc. buyout 101
department leaders 77
DFDS Dan Transport 136
DHL 133
digitizing M&A 103, 117–28
direct investments 92
distributive justice 23
dos and don'ts 9, 81

Dole Food Company Inc. 101
domestic banks 115–16
DSV 25, 39, 44, 56, 58, 70, 83, 84, 133–40, 134t, 139–40t
   M&A complementarity 59f
due diligence 111, 211
Dutch Development Bank 114

E

earnings before interest and taxes (EBIT) 188, 196
Ebix Inc. 101
Edelman's Trust Barometer 209
Elgersma, Erik xi, 41
embedded functions of corporate activity 11
emotional involvement 91, 206
employees
   ambitious 4, 10
   anxiety 126
   customer's 122
   involvement of 129
   needs and uncertainties 126
   turnover 57, 127
   unfair treatment of 127
engagement letter 100–3, 112
Engro Foods 114
enterprise value (EV) 213, 219
EPS 175, 222
exclusivity deals 109
expected outcome 7, 22
external advisors 7, 210–11

F

FAANG stocks (Facebook, Apple, Amazon, Netflix, and Google) 15–16

Facebook 15–16, 33, 74
fail fast 205, 224
failure, avoiding 2
failure rate
   of corporate M&A 4, 5, 7, 8, 10, 16, 24, 31–2, 37, 45, 46–7, 106, 182, 212–13, 217–20
   50% rule 1, 2, 11, 13, 17, 27, 31, 48, 77, 186
   global, for SMEs and large corporates 27
family owned entities 203
fast-moving consumer goods (FMCG) companies 160
fear
   about jobs and careers 126
   of M&A 80
'fear index' 14–15
feasibility 86
fees 6, 95, 102, 115
   break-up 109
   success 12, 31, 96, 102, 108, 210, 213, 220, 224
   tail 102, 112
finance department 76
financial advisors 95, 98, 99, 100, 102, 106, 209, 211–15
financial crisis, global (2008) 15, 87
Financial Economic school 19, 20, 26
'financial sponsors' 12
financials 85–6
flipping a coin 1, 5, 11, 22, 24, 48, 186, 213
FON-A-PAL 209–10
foundation for M&A 9, 63, 65, 71–3, 80

four schools of thought in M&A research 21–2t
Frans Maas 136, 137
Friesland Foods 57, 175, 176, 178, 179
FrieslandCampina 1, 4–5, 25, 57, 58, 61f, 66, 87, 99, 113–16, 220
'from push to pull' 178, 181
FT.com 104–5
FTSE 104
funding 39, 86–8, 95, 111–16, 196

G

Gaw, Kathryn xii, 25
General Electric 166
Gilead Sciences 34
Gladwell, Malcolm: *Blink* 58
global car firm 41–2
global diversification 9
Global M&A Elite 16, 17, 20–2, 23t, 225
   questionnaire to corporate executives 82–3f
   TRA 34
'Go' targets 37, 67
'go'/'no go' decisions 67, 96
goals
   achieving life's 24
   company 36–7, 42, 80, 120, 127
   legal team 105
   M&A 84, 103, 109, 223
   professional 11
'Goldman Algorithm' 58, 62, 63
'Goldman Criteria' 58
Goldman Gates 36, 37, 38, 45, 51, 89, 91, 96, 119
   definition 58–63
   origin of 62–3

   target list based on 63–8
Goldman Gates Scoring 62, 64f, 79, 81–8, 82–3f, 207
Goldman Sachs 58, 101, 115, 213
Goldman, Lee 58
good failure 48, 205, 224
Google 33, 34, 49
Google Docs 123
governance in corporate processes 95–6
government regulations 104
Green Bond ('Schuldschein') 114
growing your business with M&A 36–8

H

Handelsbanken 173, 174
handle with care projects 79
Hart, Cees 't xi, 25, 177, 178, 179, 181–2
Harvard Business Review (HBR) 169
HDFC Bank 34
hedge funds 12, 14
Heidrick & Struggles: *Accelerating performance* 33
Heineken xi, 25
Hermes 150
Hewlett-Packard Enterprise 120
historical performance 16, 20
'hold' target 67
hostile takeovers 85
human resources (HR) 73, 105, 214
'hygiene factors' 213

I

Illumina 34
indirect investments 92
information 33, 66, 73–5, 80, 81, 95, 110, 128

In-house expertise 80
integrity due diligence (IDD) 110
intent-based leadership 88–9
Interactional justice 127
Intercontinental Exchange 34
internal competencies 80
'intrinsic value' 158
investment banks 85, 108, 115–16, 202, 211–15
'investor darlings' 223

J
J. Heinz Company buyout 101
Jacobs, Marc 150, 151
Jensen, Mads xi
JP Morgan 115
judgment calls 62

K
Kaizen methodology 166
key activities 56, 57
key partners 57
key resources 56–7
King, David R. xi, 25, 27
KKR & Co L.P. 101
KKR Financial Holdings LLC 101
knowledge sharing 104
know-your-client (KYC) directive 211
Kodak 35
Koster, Marc xi, 25
Kuehne & Nagel 133

L
Larsen, Kurt xi, 25, 137
leadership
  bad 77, 88, 137–6
  and communication 119–20
  create a strong foundation for your people 71–3
  drive hard with soft management tools 70–71
  and role models 44
  silent people are not team players 73–5
  strong 2
learning organization 225
legal advisors 7, 96, 99, 102, 105, 108–9, 110, 210, 211
legal department 3, 96
letter of intent (LOI) 93, 94, 95, 109–11
LinnCo LLC 101
listening 66, 73, 209
  to the naysayers 46
  to people on the front line 46
litigation risk 101
'locked box' mechanism 110–11
Loro Piana 151, 152
LVMH Moët Hennessy-Louis Vuitton (LVMH) 15, 55, 56, 79, 149–56, 155–6t
  Business Model-Driven M&A 151–6
luck 2, 7, 11, 47, 143, 170
Lund, Jens 216

M
M&A advisors 6, 31, 87, 97, 99, 100, 102, 112, 115
M&A aspirations 4
M&A 'beauty parade' 111
M&A buy-side catalog 96
M&A Dashboard 123–4, 125t
M&A deal archetypes 65, 159, 159t
M&A Deal Committee 68, 70, 75–88, 110, 114–16

M&A Deal Types 159–60
M&A Drivers 57–8, 156–65
'M&A Event Studies' 14
M&A Formula 2, 3f
　three steps 3, 4
　applying 41–3
M&A governance 96
M&A insurance 92, 107–8
M&A Launchpad 37–8, 45, 81, 91, 92, 103–4
M&A Legal 104–11
'M&A love letter' *see* letter of intent
M&A myth-busting 217–20
M&A panel 113, 123
M&A performance miracle 18–19t
M&A Playbook 93–4, 93f, 120, 159
M&A rainmaker 49
M&A reputation 86, 201–2, 203–6
M&A stages, early
　things you should do during 205
　things you should not discuss during 204–5
M&A target list 36, 63–8
'M&A traffic lights' 95
Magnan, August 166
main processes 2
Malkiel, Burton: *A Random Walk Down Wall Street* 48
management skills 56
manufacturing footprint programs (MFPs) 196
mapping workflows 120
Marc Jacobs boutique 150, 151
market knowledge 56
market reaction to M&A deals 9

market share 63, 160, 168, 189, 191
Marquet, David xi, 88–9
material adverse changes (MACs) 47
Matzler, Kurt xi, 25, 27
'maybe' targets 37
McKinsey 9, 170, 179, 211
Mead Johnson 20
media coverage 214
Mediclinic International 106–7
memorandum of understanding 93
Mercedes 54
'mergers of equals' 174
Merton, Robert 32
metrics 25, 28, 39, 45, 120, 224
Midaxo 120–2
milk price 5, 221
'monkey business' 48
monthly retainer 97, 112
Morgan Stanley 115
Multi Development Banks (MDB) funding 114

N

Netflix 33
network-based organization 65
newspapers 48, 217
Nippon Express 133
'no go' targets 37, 67
Nokia 35
non-binding offer 94, 95
Non-Disclosure Agreements (NDAs) 113
non-organic growth 71
non-solicitation 109–10
Nurofen pill 144, 145

Index

O

onboarding times 112
openness 74, 75, 119, 188
operational excellence 44
organic growth 15–16, 33, 142, 156, 188, 210
organizational behavior 20
organizational justice theory (Greenberg) 127
organizational learning 45
Osterwalder, Alexander: *Value Proposition Design* 141
ownership, taking 2

P

'Pareto optimality' approach 97
parking space projects 68
partnership 36, 57, 78–9, 114, 203, 205, 213
patents 104
pattern of M&A 1
pension liabilities 110
performance, mediocre 5
Philips 120
planning, meticulous 11, 158
plug and play M&A IT platform 118
poor M&A practice 47, 91–2
post-merger 21, 174
preferred purchase price mechanism 111
pre-merger 21
prices 102, 162
    capital market 16
    comparison 66
    material 13
    milk 5, 221

    purchase 111, 219
    share 14, 15, 214
    stock 67, 202
pride in M&A 24
principal-agent-theory problem 206
private equity 12, 14, 107–8, 170, 203
pro-active offensive strategy 51
procedural justice 127
process school 20
procurement 76
product design 56
product health impact 110
product market attributes 55
product technology 56
professional support system 10
property interests 104
Provinsbanken 143
PTC 120
punch above organizational weight 11, 71, 214

R

Radiometer acquisition 168
RB 20, 141–6, 142t
    acquired Boots Healthcare Division 145
    timeline of acquisitions 145
real world examples 2, 4, 11, 58, 69
'Really Bad' deals 91, 94
reason for M&A 4–6
Reckitt Benckiser *see* RB
regulation 105
Rehm, Werner (and Andy West):'Managing the market's reaction to M&A deals' 9
reimbursement clauses 103

request for proposal (RFP) 95
research period 14–15, 16, 20, 33, 167
resistance
  CEO 77
  employee 125
resource attributes 55
resource library 208, 209
resources 7, 35, 56–7, 63, 77, 78, 86, 103, 115, 118, 143
respect 67, 127, 150, 209
responsibility in M&A 58, 69, 89, 100, 190
return on corporate equity (RoCE) 190
investment (return on ROI) 74
return on net assets (RONA) 158
Rimowa 150
risk 8
Risk Off 9
Risk On 9
Robinson, C.H. 133
role model, leadership and 44
Roosevelt, President 202
royalties 96
rules-based system 58, 62, 65, 94, 96

S

S&P500 Index options 14
Secher, Rikke Zink 8, 9
Sales Channels 55
Samson 136
Schenker 133
school of thought 16, 26
Schriber, Svante xi, 25, 27
Schuldschein 114
Schultz, John xi, 25, 99
securitization 105
selection bias 206–7

self confidence in M&A 214
self-fulfilling prophecy 32, 80
sell side advisors 22, 210, 219
sell-side 1
'sense of belonging' 73
Sephora 153
serial acquirers 15, 94, 149
shareholder alignment 221
shareholder value 2, 5, 31, 36, 47, 94, 103, 136, 147, 206, 208
Shire 34
'should be' business model 70
single deals 216–17
Sinotrans 133
Slack 74
Slaughter and May 106–7
SMA Research Lab 127
small and medium enterprises *see* SMEs
SMEs 1, 4, 6, 10, 14, 16, 17, 22, 32, 44, 45, 91, 105, 106, 108, 212, 214–15, 218
  16 questions 26, 51, 53, 58
  survey 48, 51–3, 52f, 58
  vs large multinationals 26–7
social media 124, 209
Softbank Group 34
software 103, 118, 127, 223
speed of execution 105, 111
Springer, Klaas xi, 25, 113–16
SSL 145
Standard Operational Procedure (SOP) 95
Starboard Cruise 153
Starbucks 34
statistics 11, 26, 57–8, 100, 157, 214, 222
stereotype of M&A deal 217

Index

strategic collaboration 105
'strategic fit' 19
Strategic Management Journal 16, 17
'Strategic Management' school 19
Strepsils brand 141, 144, 145
stretch goals 36, 38, 99, 211
students 1
success 2, 46–9, 222–3, 224–5
    antecedents of 17, 51–3
    deals you don't do 6, 77
    rate 1, 11, 75, 112, 213
    three rules of 4, 32
success fees 12, 31, 96, 102, 108, 210, 213, 220, 225
super-accelerators 33
suppliers 56
supply channel types 56
survival period of liabilities 107
sustainability 110
Swiss Central Bank 47

T

tail fees 102, 112
taking ownership in M&A 44–5, 91–116, 120, 224
target list 63–8
tax 3, 70, 76, 107, 208, 211
    department 96
    optimization tools 96
teamwork 165–73
TechCrunch 54
technical skills 56
technology 12, 33, 35, 54–5, 57, 74, 105, 188, 190, 191, 194t, 195, 196, 209–10
Tellabs Inc. buyout 101
termination rights 102
Terms and Conditions in M&A 102
Tesla 33
theory into practice 24
third-party logistics (3PL) firms 133–40
Thomas, Patrick 150
thought leaders 77
Three-Step M&A Formula 38–40
    Follow Business Model-Driven M&A. 38–9
    Exercise Strong Leadership and Communication 39
    Take Ownership 39–40
Tolstoy, Leo 79
Toray Industries Inc. 101
total return analysis (TRA) 15, 16, 19, 20, 25, 26, 35
trading multiples 66, 136
transaction fee (aka 'success fee') 102
transaction multiples 66
transaction service providers (TSPs) 99
transformative decision 32
transformative M&A 11
transparency 46, 70, 74, 75, 119, 120, 127, 159, 170, 182, 215, 218
'travelling in time' with corporate M&A 55
Treasury 76
    role of 112–16
trial and error 4
true growth 16
Tullberg, Leif 134

U

Uber 54
uncertainty on an M&A 10, 12, 73, 160

unidimensional measure 18
UPS 133
US Navy 88
UTI 136

V

valuation 66, 78, 109, 122, 169, 191, 217
  myth 218
  standards 18
value 8
  adding 31, 42, 56, 137, 153, 158
  creation 12, 27, 41, 44, 57, 107, 137, 157, 166, 169, 172, 176, 187, 189
  client 104, 120
  enterprise 213, 219
  intrinsic 158
  market 17, 20, 26, 158
  shareholder 2, 5, 31, 36, 47, 94, 103, 136, 147, 206, 208
  of staff 124
value chain 55, 168
value proposition 46, 56, 76, 84, 141–8, 150–1, 153, 162, 184–5, 191, 194, 220
Verizon 25
VISA 34, 35
VIX 14–15

Volkswagen 54
VUCA (volatility, uncertainty, complexity, and ambiguity of general conditions and situations) 13

W

Wall Street Journal 48
weighted average cost of capital (WACC)/economic value added 158, 210, 219
Wiley's Online Library 26
Williams, Serena 9–10, 92, 93, 110, 186
winning mindset 10, 39
working capital, net 111
World Bank 114
World's Top 10 Investors 8, 10, 98

X

Xerox 35

Y

Yale 13

Z

Zoltek Company Inc. 101